International Political Economy Series

General Editor: **Timothy M. Shaw**, Professor and Director, Institute of International Relations, The University of the West Indies, Trinidad & Tobago

Titles include:

Dick Beason and Jason James
THE POLITICAL ECONOMY OF JAPANESE FINANCIAL MARKETS
Myths versus Reality

Mark Beeson
COMPETING CAPITALISMS
Australia, Japan and Economic Competition in the Asia-Pacific

Shaun Breslin
CHINA AND THE GLOBAL POLITICAL ECONOMY

Kenneth D. Bush
THE INTRA-GROUP DIMENSIONS OF ETHNIC CONFLICT IN SRI LANKA
Learning to Read between the Lines

Kevin G. Cai
THE POLITICAL ECONOMY OF EAST ASIA
Regional and National Dimensions
THE POLITICS OF ECONOMIC REGIONALISM
Explaining Regional Economic Integration in East Asia

Gregory T. Chin
CHINA'S AUTOMOTIVE MODERNIZATION
The Party-State and Multinational Corporations

Yin-wah Chu (*editor*)
CHINESE CAPITALISMS
Historical Emergence and Political Implications

Abdul Rahman Embong
STATE-LED MODERNIZATION AND THE NEW MIDDLE CLASS IN MALAYSIA

Takashi Inoguchi
GLOBAL CHANGE
A Japanese Perspective

Dominic Kelly
JAPAN AND THE RECONSTRUCTION OF EAST ASIA

L. H. M. Ling
POSTCOLONIAL INTERNATIONAL RELATIONS
Conquest and Desire between Asia and the West

Pierre P. Lizée
PEACE, POWER AND RESISTANCE IN CAMBODIA
Global Governance and the Failure of International Conflict Resolution

S. Javed Maswood
JAPAN IN CRISIS

International Political Economy Series
Series Standing Order ISBN 978-0-333-71708-0 hardcover
Series Standing Order ISBN 978-0-333-71110-1 paperback
(*outside North America only*)

You can receive future titles in this series as they are published by placing a standing order.
Please contact your bookseller or, in case of difficulty, write to us at the address below with
your name and address, the title of the series and one of the ISBNs quoted above.

Customer Services Department, Macmillan Distribution Ltd, Houndmills, Basingstoke,
Hampshire RG21 6XS, England

Local Climate Governance in China

Hybrid Actors and Market Mechanisms

Miriam Schröder
Research Fellow, Potsdam University, Germany

First published 2012 by
PALGRAVE MACMILLAN

Palgrave Macmillan in the UK is an imprint of Macmillan Publishers Limited, registered in England, company number 785998, of Houndmills, Basingstoke, Hampshire RG21 6XS.

Palgrave Macmillan in the US is a division of St Martin's Press LLC, 175 Fifth Avenue, New York, NY 10010.

Palgrave Macmillan is the global academic imprint of the above companies and has companies and representatives throughout the world.

Palgrave® and Macmillan® are registered trademarks in the United States, the United Kingdom, Europe and other countries.

ISBN 978-0-230-30161-0

This book is printed on paper suitable for recycling and made from fully managed and sustained forest sources. Logging, pulping and manufacturing processes are expected to conform to the environmental regulations of the country of origin.

A catalogue record for this book is available from the British Library.

A catalog record for this book is available from the Library of Congress.

10 9 8 7 6 5 4 3 2 1
21 20 19 18 17 16 15 14 13 12

Contents

Tables

Illustrations

Map

Figures

Abbreviations

ACCA 21	Administrative Center for China's Agenda 21
ADB	Asian Development Bank
AFD	Agence Française de Développement
BDS centers	Business development service centers
CCB	Climate, Community and Biodiversity (standards)
CCICED	China Council on International Cooperation for Environment and Development
CDM	Clean Development Mechanism
CER	Certified emission reduction
CIC	China Investment Corporation
CIDA	Canadian International Development Agency
CNY	Chinese yuan
DFAIT	(Canadian) Department of Foreign Affairs and International Trade
DNA	Designated national authority
DNV	Det Norske Veritas
DOE	Designated operational entity
DRC	Development and Reform Commission
EB	Executive board
EPB	Environmental Protection Bureau
ERPA	Emission Reduction Purchase Agreement
EU ETS	European Union Emission Trading Scheme
FDI	Foreign direct investment
GDP	Gross domestic product
GEI	Global Environmental Institute
GHG	Greenhouse gas
GIZ	German International Cooperation
HFC	Hydrofluorocarbon
IMET	Italian Ministry for Environment and Territory
IPCC	Intergovernmental Panel on Climate Change
IRR	Internal rate of return
KIBS	Knowledge-intensive business services
KISA	Knowledge-intensive service activities
MDG	Millennium Development Goal
MEP	Ministry of Environmental Protection
MFO	Market facilitation organization

MOFCOM	Ministry of Commerce
MOST	Ministry of Science and Technology
NDRC	National Development and Reform Commission
NEPA	National Environmental Protection Agency
NGO	Non-governmental organization
NIE	New institutional economics
NLGCC	National Leading Group on Climate Change
ODA	Official development assistance
OECD	Organisation for Economic Co-operation and Development
PDD	Project design document
PFC	Perfluorocarbon
PIN	Project idea note
PoA	Programme of activities
R&D	Research and development
REDD	Reducing Emissions from Deforestation and Forest Degradation
SD	Sustainable development
SMIDA	Small-industry development organization
SOE	State-owned enterprise
STD	Science and Technology Department
TNC	The Nature Conservancy
TVE	Township and village enterprise
UNCTAD	United Nations Conference on Trade and Development
UNEP	United Nations Environmental Programme
UNFCCC	United Nations Framework Convention on Climate Change
USD	US dollars
VER	Voluntary emission reduction
WB	World Bank
WTO	World Trade Organization

Acknowledgments

First of all, I would like to convey my sincere thanks to all my interview partners in China who gave their valuable time to participate in the extensive interviews for this book. Most of the interviewees for this research were very patient and open minded, and I hope I have managed to convey their personal impressions and evaluations of the Chinese carbon market in this book. Readers will understand that publishing the names and affiliations of the interviewees would have compromised their anonymity and thus would have made an open exchange of ideas less possible.

I am greatly indebted to Dr Andreas Oberheitmann and Dr Wang Can, directors of the Research Center for International Environmental Policy, Tsinghua University, who invited me as a guest researcher for five months between September 2007 and February 2008. This research would never have been possible if they had not provided me access to their networks, and thereby opened doors in China to meet people who would not otherwise have been available. Special thanks also go to Mr Li Gao, formerly with MOST, now director of the Climate Change Department, National Development and Reform Commission (NDRC), who facilitated access to the four CDM centers. Very special thanks also go to the directors of the four CDM centers, who were willing to share their hands-on experiences of CDM capacity building and project development.

I would also like to express my gratitude to Professor Harald Fuhr (Chair of International Politics, Potsdam University), Professor Miranda Schreurs (Director of the Environmental Policy Research Centre, Free University Berlin), and Dr Axel Michaelowa (Head of Research, 'International Climate Policy', University of Zurich), all of whom provided constructive comments on this research and the making of this book. Attending the annual staff workshops at the SFB, the biannual retreats of our Potsdam University project team and the bimonthly research meetings on questions of environmental governance at the Free University Berlin made continuous discussion with other experts possible.

Many thanks go also to my colleagues at the SFB 700 (www.sfb-governance.de), who provided both substantive criticism and help with getting this book into shape: Dr Markus Lederer, Dr Marianne Beisheim,

Professor Tanja Börzel, Gudrun Benecke, Lars Friberg, Melanie Müller, Philipp Binias, Hendrikje Reich and many other SFB researchers.

I would also like to express my gratitude to the UNEP Risø Center, Denmark, for their permission to print some figures from their popular CDM Pipeline (www.cdmpipeline.org).

Introduction

The local dimension of global climate governance

Climate change has a clear global dimension, and public attention is naturally focused on international agreements as a possible solution. The Kyoto Protocol did indeed provide a successful international regime to commit industrialized countries to greenhouse gas (GHG) emission cuts. Lately, however, international climate politics has reached a stalemate, and the multilateral climate negotiations within the United Nations framework have achieved frustratingly little. In the face of a stalemate in international climate politics, attention is refocusing on multilevel governance and climate initiatives at the local level (Gupta, 2009; Schreurs, 2008; Sippel and Jenssen, 2010). Federal state action in the United States (Lutsey and Sperling, 2008) and elsewhere (for example, South Africa and Japan – Roberts, 2008; Sugiyama and Takeuchi, 2008) and city-level climate programs are emerging (Corfee-Morlot et al., 2009), not so much as replacements, but hopefully as a complements to any future international climate regime. Local-level climate action is indeed a necessary bottom-up process that can support and complement the top-down approach of international climate politics. The two approaches go hand-in-hand: civil societies' demands urge national governments to commit to international climate cooperation, while climate targets agreed upon at the international level can only be successfully implemented domestically if local actors are effectively engaged.

This book investigates how local climate capacities have been built at the provincial level in China in order to contribute effectively to the international clean development mechanism (CDM). This approach to analyzing the local capacities and political will necessary for climate

1

protection complements today's prevailing discussion about national reduction targets and international climate agreements. Especially when it comes to China, the debate turns on its future global position as the world's largest GHG emitter and its alleged obligation to take up fierce reduction commitments internationally. What is often left out of the discussion is the focus on local capabilities and incentives that are required to implement such international commitments on the ground. The example of the CDM is interesting: the market mechanism has been designed and put into force at the international level, but its effective implementation depends heavily on the political support of local politicians, the capacities of local project developers, and the willingness of local banks to provide the necessary upfront financing. Despite a transitional economy that certainly does not provide ideal textbook conditions for the handling of market mechanisms, China has become the world's leading CDM host country. The surprising success story of the CDM in China demonstrates that market mechanisms like the CDM can be made to contribute to emission reductions even in countries with transitional economies, where market actors are public-private hybrids. At the birth of the Chinese carbon market, in 2005, local companies, government officials, and banks were very skeptical about the idea of trading emission certificates internationally, and refrained from participating. Only two years later, in 2007, China became the world's number-one CDM host country, and still accounts for the majority of CDM projects and certificates in the global market (UNEP Risoe, 2010). This book's intention is to shed light on one explanation for the Chinese CDM success story, namely its local market catalysts – the so-called provincial CDM centers – and their role in facilitating the diffusion of the CDM at the local level.

Improving China's national system of environmental governance

In the world's inglorious ranking of top polluters, China is quickly catching up. In 2007, China at last overtook the US as the world's largest emitter of CO_2 (Netherlands Environmental Assessment Agency, 2008); it is also tops lists relating to the world's most heavily polluted cities and severest local air and water pollution (World Bank, 1997a; OECD, 2001; Economy, 2004). At the same time, China overtook the US as the world's heaviest energy user in 2009, and is projected to account for 22 percent of the world's energy demand by 2035 (IEA, 2010b: 47). As home to the world's largest national population – whose per-capita energy consumption remains low but is set on a paths towards high economic growth and improved living standards – China's demand for both energy and natural resources is growing sharply. China's surging demand threatens to further unbalance the earth's carrying capacity, which has already been shaken tremendously by the consumption patterns of industrialized countries. The unconditional pledges for GHG emission reductions made by China and other countries in the aftermath of Copenhagen contribute only 60 percent of the necessary reductions needed to stabilize global warming at around 2 degrees Celsius (UNEP, 2010). Seen from the perspective of a global carbon budget that needs to be limited at around 450 parts per million, emissions from China need to peak in 2020 and then be curbed as soon as possible (IEA, 2010a: 383). The achievement of China's emissions peak between 2030 and 2040, as is predicted by Wan Gang, China's science and technology minister, is likely to be too late (Watts, 2009).

Consequently, international pressure on China to act is increasing. The Western media have accused China of being the global climate villain in the aftermath of the Copenhagen conference (Watts et al., 2009; Vidal, 2009), although some good will was attributed to China's performance in Cancún (Chipman and Morales, 2010). The US and Japan have made their participation in any new UN-backed climate regime conditional on Chinese commitments. European and American suggestions of imposing border-adjustment measures on carbon-intensive import products make climate-related trade wars an unwelcome possibility (Economist, 2009; Reuters, 2009). Not surprisingly, anger is mounting in China and other emerging economies at their being identified as the culprits, thereby justifying Western countries' unwillingness to transform into low-carbon economies. In contrast to the Western perspective, the Chinese media portray China's role in the climate negotiations as constructive and progressive (Fu and Li, 2009). Chinese diplomats at the UN negotiations do not tire of reminding the developed world of its (often unfulfilled) emission-reduction commitments; they press for the continuation of the Kyoto Protocol and demand a global burden-sharing agreement that will respect developing countries' right to development. The international debate about burden-sharing is intense when it comes to the questions of which countries must take responsibility, which must take action, which must pay for such action, and which must endure the negative consequences of global inaction. In a nutshell, the focus is on deciding *who* is to provide for the global public good of climate protection. At the national and local levels, however, the question of *how* to provide for climate protection is of utmost importance. Once China has agreed to GHG mitigation targets (for example, to reduce the carbon intensity of its GDP by 40 to 45 percent, as it pledged in January 2010 – NDRC, 2010), the challenge will move from the international negotiation stage to the national, and ultimately local political stage, where these international obligations must be translated into domestic politics.

The next questions looming on the horizon of international climate debate therefore relate to local policy implementation. Once the Chinese government has agreed upon climate targets internationally (whatever scope these targets may have), the challenge is to come up with adequate measures for achieving these targets within China – especially at the local level, where climate politics are not yet part of the political agenda and environmental politics generally are of less priority than aspirations for economic growth. The future challenge for China thus turns on the question of how to provide for climate protection at the local level.

The question of how to curb GHG emissions in the world's largest emitter has in many ways become a question of how to improve China's national system of environmental governance (Mol and Carter, 2007). 'Governance' usually refers generally to all forms of collaboration between state actors and private organizations – both profit-making and non-profit – for the achievement of public objectives (Risse and Lehmkuhl, 2006: 7, Biermann and Pattberg, 2008: 280); while environmental governance implies a special focus on 'the approaches used by a society to address pollution and promote conservation (Schreurs, 2006: 57).' The history of environmental governance in China reveals that setting ambitious targets at the central level is not enough, and can be seen only as the first step in the tortuous process of inducing behavioral change on the part of polluters at the local level. Although issues like the deteriorating state of the environment and the negative impacts of climate change are climbing up the political agenda in China, traditional state-imposed command-and-control instruments have done little to improve things. For example, targets on SO_2 emission reductions could not be reached as long as they were pursued by imposing levies on factories that were barely enforced by the local environmental protection bureaus (Tang et al., 1997). However, facing a looming environmental catastrophe at home and international pressure to contribute to the global fight against climate change, fundamental shifts are beginning to take place in China's system of environmental governance (Ma and Ortolano, 2000; Harris, 2005; Mol and Carter, 2006; Ho and Vermeer, 2006). The match between centrally determined policy targets and effective local policy implementation can thus be identified as one of the greatest challenges to China's system of environmental governance.

With its economy in transition, China is in the midst of dismantling the planned economy system, pampering the private sector and experimenting with market approaches (Naughton, 1995; Chu, 2010). This has consequences for how environmental issues are dealt with, because China is also revising its environmental governance system by taking up market-based instruments (Shuwen, 2004; Economy calls this process 'economisation of environmental governance' – Economy, 2007: 172). In OECD countries, market-based mechanisms became fashionable in the 1990s because they at least put a price-tag on natural resources. Market mechanisms have also been increasingly recommended for developing countries in general (Panayotou, 1998; World Bank, 1996; World Bank, 1997b) and for China in particular (ADB, 2001). They usually involve a competitive market with individual buyers and sellers transacting on a voluntary basis (Grieg-Gran et al., 2005: 1,512). Market mechanisms or market-based instruments can be defined as 'regulations that encourage behavior through market signals rather than through explicit directives regarding pollution control levels or methods (Stavins, 2001: 1).' Although the national-level introduction of market mechanisms is still rare, they are experimented with within a broad range of sectors and local contexts. Two central questions thus present themselves: Is the introduction of market mechanisms a viable alternative to the state-centric governance approach in China? And what can we learn from China's policy experiments with market mechanisms about how to improve its environmental governance?

The CDM success story

The CDM is one market mechanism that has so far been a great success story in China: CDM projects have reduced GHG emissions on a large scale; they have generated extra revenue for government and profits for private companies, and they have contributed to job-creation and increased awareness of environmental issues. The success of this market mechanism supports arguments that increasing the number of market-based approaches in environmental governance might be effective even in a transitional country like China.

The CDM is an international market mechanism established by the member-states that are parties to the Kyoto Protocol. It has a dual objective: to achieve both cost-efficient GHG emission reductions and contributions to sustainable development. The principle behind the mechanism is that, in order to stop global warming, it does not matter where GHG emissions are reduced. Because GHG emission reductions tend to be cheaper in developing countries, governments and companies pay these countries for emission reductions through a system of certificate-trading, rather than by reducing emissions at home. Accordingly, CDM projects in developing countries reduce GHG emissions, for which 'certified emission reductions' (CERs) are issued. These certificates are tradable on the global carbon market, and countries with emission-reduction targets under the Kyoto Protocol can make use of CERs to meet a certain percentage of their GHG reduction obligations. The host countries of the CDM receive the benefit of monetary compensation for the costs of GHG emission abatement, and are also able to reap additional sustainable development benefits due to the CDM projects. Thus, contrary to the traditional command-and-control regulations of environmental governance, the CDM offers monetary benefits as an incentive for environmentally friendly behavior. This represents an innovation in China's approach to environmental governance.

One important reason for the CDM's success in China is its enormous national potential for reducing GHG emissions. In addition, carbon market brokers often point to the good investment conditions China provides. Is carbon finance thus flowing to China for the same reasons as any other foreign direct investments (FDI)? Part of the answer must be yes. But investors also complain about the strong interference of the Chinese government in the carbon market: it has placed strict ownership restrictions on foreign companies, limiting their stakes in CDM projects to 49 percent and imposing a tax of up to 65 percent on profits on certain types of CDM projects, and has shown an open preference for Chinese companies, amounting in effect to protectionism. Such strong state interference in a market mechanism is typical of China's governance style, which combines the use of market mechanisms with the retention of government control (Zhang, 2002). Such a deployment of market mechanisms for the achievement of political objectives can already be observed in CDM governance at the national level in China (Schroeder, 2009). How does the state-market relationship within CDM governance look at the local level? Although research on this topic is so far scarce (it focuses on CDM project implementation – see, for example Zhang et al., 2005), initial investigations seem to confirm strong state intervention in the CDM at the local level in China (Qi et al., 2008: 394).

By taking a closer look at how public and private actors – and their hybrids – become involved with the CDM at the local level, this book discusses how, in the aggregate, the introduction of market mechanisms changes the relation between state and market in environmental governance in China – for both better and worse. Using this approach, the book provides a first assessment of recent policy experiments with the introduction of market mechanisms in China's environmental governance system: it investigates how local climate capacities had to be built and encouraged at the provincial level in China in order to contribute effectively to the international CDM. The surprising success story of the CDM in China highlights the means by which such market mechanisms like the CDM can be made to achieve emission reductions even in countries whose economies are in transition, and with market actors that are public-private hybrids.

Case study on the performance of provincial CDM centers as hybrid actors

It is often forgotten that China had not always been the frontrunner in the global CDM market. On the contrary, China initially lagged behind other CDM host countries because the CDM faced several barriers in China. These included the skeptical attitude of the Chinese government towards the CDM, a lack of trust in the mechanism on the part of companies, and an absence of capacities and financial means to develop CDM projects at the local level (see Chapter 2). The saying that 'the CDM is like a cake falling from the sky,' which was very common in China in 2005 and 2006, betrays the general disbelief in the CDM at the time. The CDM as an institution was not known or trusted, and potential market participants were unable to evaluate its benefits and risks. Being traditionally rather risk-adverse, most local companies thus refrained from taking the CDM into consideration when it came to project financing. The early CDM market in China thus experienced a market failure due to a lack of adequate information about the CDM as a means of project finance, as well as an absence of knowledge about project eligibility and experience of how to plan, implement and monitor CDM projects. Providing these necessities was too costly for private companies, but was necessary in order to kick-start the market. This was especially true for China's western provinces, which have good CDM potential but were lagging behind in using the CDM because they lacked the relevant knowledge and experience. These barriers to the CDM market needed to be overcome.

Because the CDM market in China did not evolve quickly, and at first showed signs of market failure resulting from asymmetric information, the Chinese government decided to support the initiatives of some local government officials to establish so-called 'provincial CDM centers' in order to kick-start local CDM markets. They did this in line with previous Chinese policy measures supporting innovation and new markets. For example, the setting up of technology-promotion centers tasked with diffusing new technologies is common in China (Jefferson, 2006). Within a political system relying more on personal relations and the influence of leading political figures than on laws or regulations (Lieberthal and Oksenberg, 1988), the approach of setting up 'centers' to push political ideas is typical. Although the first idea for a CDM center came from an individual from Ningxia, the central Chinese government soon took up the idea and approached foreign donors for financial assistance. Twenty-seven of these centers now exist or are in the process of being established. Most are semi-public, semi-private agencies with the dual mandate of both providing public goods and operating at a profit. Their tasks include CDM promotion, capacity development, and, last but not least, CDM project development. Despite their similar mandates, the centers show a large variation in their effectiveness in achieving their objectives. Some centers have achieved monopolies in their provincial CDM market, while others are not even notable market players. Taking the numbers of CDM projects developed as an index of their effectiveness, nine centers were able to develop projects that reached validation stage, while only three centers were able to contribute 10 percent or more of their province's share of CDM projects (see also Table 2.3).

Why do these centers vary so markedly in their performance? Taking a closer look at their performance is worthwhile in two ways: first, it can contribute new insights about the pros and cons of state intervention for diffusing green products and facilitating the emergence of markets; second, these CDM centers are the first of their kind in the world, so that identifying factors in their success will be helpful in establishing similar approaches in other CDM host countries. This book's intention is to shed light on the role of market catalysts in facilitating the diffusion of the CDM at the local level. The question addressed by the detailed research for this book is thus: What factors explain the effectiveness of CDM centers as hybrid agencies in diffusing the CDM and facilitating local CDM markets?

At the systemic level, the CDM centers are a good example of how the state is changing its role in relation to new market forces. The empirical findings in this book reveal that CDM centers do not fit into the public-private dichotomy, but are hybrid actors with varying degrees of attachment to the government. They are thus an example of the growing role of hybrid actors in environmental governance in China; they are hybrid in nature because they have to adhere to the public mandate to provide the public goods of CDM information, knowhow, and experience while also operating on a for-profit, businesslike basis. These kinds of hybrid actors are a widespread phenomenon in China. On the one hand, they are consequences of the Communist legacy of state ownership; on the other, they are products of the downsizing and outsourcing of tasks formerly carried out by a now revenue-stripped public sector. There is a widespread expectation that they will be able to supplement the weak local state in environmental governance, because they are supposed to be more effective and efficient than their public counterparts. There are, however, also doubts about how far they can be compelled to provide public goods if they grow economically independent of their public patrons. An examination of the CDM centers for their effectiveness can thus reveal much about the role and performance of hybrid actors in China's environmental governance. The broader research question addressed by this book is thus: How can we explain the emergence and performance of hybrid actors like the CDM centers in environmental governance in China?

This book adopts a two-fold approach for tackling these two central research questions. Firstly, a classical comparative case study is used to examine the performance of the CDM centers as semi-public agencies for market facilitation. Secondly, lessons are drawn from these findings to evaluate the performance of hybrid actors in environmental governance.

Provision of public goods by hybrid actors

For the CDM to function smoothly, new market institutions had to be created within the host countries. The creation and maintenance of market institutions is often seen as a public good, because they are non-rival and non-excludable in their usage by market players (Olson, 1971; Xia, 2000: 34, Kaul et al., 2005: 21; Kaul, 2008). The establishment of CDM market institutions at the international, national, and local levels can thus also be regarded as provision of public goods for a market, which in turn is thought to provide the global public good of a stable climate. In order to set up international CDM market institutions, new property rights had to be created for the usage of air as a sink for GHG emissions. CERs were created as a tradable good certifying that one ton of CO_2e emissions had been reduced. Trade and exchange rules, such as the European Union Emissions Trading Scheme (EU ETS) linking directive were determined, and market oversight institutions such as the Executive Board (EB) and Designated National Authorities (DNAs) were established. The market institutions governing the international CDM market were created mainly on the basis of Marrakech Accords of the Kyoto Protocol. National CDM regulations on project approval and eligibility complemented the international regulations, and the setting up of national DNAs fulfilled the minimal framework conditions required by the Marrakech Accords for a country to become a CDM host country. Despite the CDM's 'invention' at the international level, it still faced many barriers at the host-country level that hampered its diffusion.

The situation of the early Chinese CDM market can be seen as typical of the diffusion of innovations at an early stage. Innovations often need some kind of private and/or public intervention in order to be kick-started. They are financed either by the innovators themselves – for example, large companies – or, where it is in the public interest to have these innovations commercialized and diffused, by public institutions. External intervention to catalyze the diffusion of innovations is often necessary, because of the paradox surrounding innovation. Even products that only have advantages are sometimes not easily diffused. The uptake of new innovations is hindered by a general tendency to oppose change. Adopters might have no knowledge of the innovation or misunderstand it; perhaps there is a reluctance to relinquish old habits, a desire to follow trends, or simply a distaste for the innovation. This hesitancy about change is a nuisance for sellers of mobile phones or software, but it is a serious problem for policy-makers hoping to introduce either innovative products, policies, or practices designed to enhance environmental protection. There are many examples for such innovations that have enjoyed public support because of their benign character: most prominent among them are renewable energies (Murphy and Edwards, 2003), energy-efficient applications (Jänicke et al., 1999), and public health insurance (Greenberg, 2006). While their use is often likely to benefit the individual, society, and the environment, these innovations often are not diffused as quickly as their proponents would wish. This situation has been called the paradox of 'high potential, but low diffusion' (Shama, 1983; Douthwaite et al., 2001). Because we need to quickly step up our efforts to safeguard the environment and to stop global warming, it is of utmost importance that we establish how – as a first step – we might at least put to use the green products, policies, and practices that might help us to overcome our societies' dependence on fossil fuels (Foxon, 2003; Grubb, 2004). Putting innovations to use is thus one important step in improving our ability to combat climate change.

The CDM can thus be seen as an innovation intended to contribute positively to environmental protection, but which nevertheless is not automatically or easily diffused. I argue in this book that the CDM represents an innovative financing mechanism for projects that reduce GHG emissions, which needs to be diffused in a similar way to other innovations before achieving full acceptance by its users. More specifically, the CDM can be seen an innovative product within the newly evolving industry of the trading of emission rights. Like other products of this industry, the CERs of CDM projects move through a product life-cycle that includes phases of invention, diffusion, and diversification. The CER was 'invented' as a tradable good on an international level, having no predecessor in the CDM host countries.

One aim of this book is thus to explore how to improve environmental governance by kick-starting local markets for green products – specifically, in relation to the provincial diffusion of the CDM in China. Indeed, public support has also proved necessary for the early development of CDM markets. It is in the interests of both the international community and the Chinese state to have a functioning national CDM market in order to achieve a cost-effective means of reducing emissions, to gain access to advanced technologies, and to contribute to (sustainable) economic development in China's underdeveloped western regions. Assuming that the CDM itself is effective in its achieving its goal (an assumption that is widely contested – for a summary such arguments see Schneider, 2007), it would be in the public interest to have a functioning CDM market. Initiated by the Chinese government, the provincial CDM centers can be seen as one form of public intervention in the local CDM diffusion process.

This book's argument relies on the debate on the changing relation between public, private, and hybrid actors in the provision of public goods in environmental governance (Rhodes, 1996; Pierre and Peters, 2000; Pattberg and Stripple, 2008). The analysis of how to initiate new markets for emission certificates within this system in transition can provide new insights, especially on the intricate relationship between public and private actors in the Chinese economy and on the difficulty of reaching a balance within the central-local nexus in both policy-making and implementation. In the case of China, an analysis of the effectiveness of multi-level environmental governance is vital, because the local implementation of fairly progressive national policies is still the main deficit (Economy, 2007; Turner and Linden, 2007; Schreurs, 2008: 351, Qi et al., 2008). Analysis of the effectiveness of provincial CDM centers should reveal new insights into how local hybrid actors perform within China's environmental governance regime for the provision of public goods.

*   *   *

The following chapters are organized according to the two-fold research approach outlined above. The detailed discussion of localized CDM diffusion and market development constitutes the core of the book. But the 'CDM story' is also embedded within a more general debate on the changing system of China's environmental governance. The two levels are interlinked because the research object of the ground-level analysis – the CDM centers – provide one example of the emerging hybrid actors in China's 'marketized' environmental governance. While the book can of course be read from the first page to the last, in order to appreciate the whole argument, readers may also choose to focus either on the ground-level discussion (Chapters 2 to 5) or on the overarching argument (Chapters 1, 6 and 7).

Chapter 1 introduces the reader to the broader relevance of this volume by outlining why China's environmental governance needs reform in order to be able to contribute to global efforts to combat climate change. The chapter first gives a brief introduction on the challenges that China represents for the international climate negotiations, and the challenges climate change poses for China. It then reviews the current Chinese paradox of having high environmental protection targets but low implementation rates, and identifies bureaucratic inefficiencies at the local level as the main reason for this. The question is addressed of whether market mechanisms might more effectively deliver environmental objectives, and several Chinese policy experiments with market mechanisms are briefly discussed, including the CDM. Such policy experiments usually incorporate considerable state intervention, which is not always successful. So the question of what conditions are conducive to state facilitation of market mechanisms is examined in relation to the empirical case of provincial CDM centers. Finally, the chapter outlines how CDM centers offer one avenue for exploring the interplay in China between public and private actors for the provision of public goods for environmental governance (specifically, information and capacity for the facilitation of green product diffusion).

Chapter 2 introduces the reader to the ground-level 'CDM story.' It starts by providing an overview of the development of the Chinese CDM market, at both national and provincial levels. The chapter identifies past and present barriers to CDM market development, and specifies provincial needs and priorities for capacity development. It gives an overview of donors' projects aimed at capacity development for the CDM at the provincial level in China, provides an analysis of the underlying interests and motives of the actors involved in setting up provincial CDM centers, and introduces the CDM centers and their objectives and functions. The chapter also outlines the puzzle of why CDM centers show such variability in their effectiveness.

Chapter 3 outlines the conceptual and analytical approach for the case study comparison. It draws on insights from innovation theory about how the diffusion of new products usually takes place, and how this process can be facilitated by public or private intervention. In addition, practical experience from organizational development of how comparable market-facilitation organizations are working in other issue areas is reviewed for success factors. From existing theory and experience, several explanatory factors are derived which guide the empirical analysis of Chapter 4.

Chapter 4 forms the empirical core of the book. The chapter summarizes the analyses of four empirical case studies of local CDM market development in Ningxia Autonomous Region, Gansu Province, Hunan Province, and Yunnan Province. Based on Mill's 'method of difference' (Mill, 1973: 388), four case studies have been selected for the empirical analysis of CDM centers. These cases show varying degrees of effectiveness, for which a proxy indicator is the number of CDM projects that have been successfully developed. Provinces are selected in such a way as to display similar framework conditions. Thus, only provinces from the western and central parts of China that are comparable in their economic development and CDM potential are taken as case studies. Chapter 4 provides a summary of the individual analyses that have been conducted for each selected CDM center and its provincial market. The summary is structured according to the potential explanatory factors for the centers' effectiveness that are identified in Chapter 3. In addition, explanatory factors are identified for which neither innovation theory nor previous practical experience with similar diffusion catalysts have created an expectation.

Chapter 5 discusses and compares the results from the case studies for their direct relevance to the CDM center approach. The first section of the chapter outlines the factors that have been found to be explanatory for the centers' effectiveness, and discusses whether these are in line with the assumptions from innovation theory or whether they go beyond it. Some crucial factors, such as organizational setting, are discussed in detail for their theoretical and practical implications. Although the CDM center approach has delivered mixed results in China, it is recommended for other CDM host countries that need to build up local CDM capacities. Some suggestions for improvements are outlined, and the general transferability of the Chinese model to other CDM host countries is evaluated.

Chapter 6 analyzes the results of the case studies for their implications for environmental governance in China. It highlights important features of the CDM centers that were neither expected nor could be explained by innovation theory – for example, their semi-public, semi-private institutional setting. Because the CDM centers resemble other hybrid actors in China, this chapter takes a broader perspective on the emergence and performance of hybrid actors, and also distinguishes between hybrid actors in OECD countries and in countries with economies in transition. The argument is advanced that CDM centers, like other hybrid actors in China, are merely precursors of an increasingly entrepreneurial approach within the Chinese governance system.

Finally, Chapter 7 discusses the implications of this marketization of environmental governance and the underperformance of hybrid actors in the provision of public goods. The question is raised of whether hybrid actors in China are only transitional phenomena, or indeed forerunners of a new Chinese 'entrepreneurial-corporatist' model of governance. Surveying some areas for future research, the book discusses whether hybrid actors can become a sustainable organizational governance structure, and what options might be available for ensuring their accountability and improving their provision of public goods for climate protection.

Part I
The Challenge

1
Climate Change as a New Challenge to Environmental Governance in China

Climate change as a global challenge

The debate on a post-2012 Kyoto regime depends to a large extent on the question of how to include developing countries into the global endeavor to cut GHG emissions. The use of the CDM has been a first step in raising revenues and capacity for GHG emission reductions in developing countries. Demands for a more stringent inclusion of developing countries in the world's efforts to combat climate change have become louder – the US has virtually made its participation in any successor to the Kyoto Protocol conditional on the inclusion of large GHG emitters such as China and India. From the perspective of most developing countries, the principle of 'common but differentiated responsibilities' ensures them the right to develop first, while it puts the main responsibility for cutting emissions on industrialized countries. But if developing countries realize their ambitious economic growth targets and their legitimate achievement of comparable living standards to those in the West without a restructuring of the energy system, there will be no possibility of mitigating climate change. Instead, various factors will conspire to produce a gloomy picture: pressure on natural resources, together with high population densities in an already deteriorated environment, will exacerbate the negative impacts of climate change, which will in turn undermine the development capacities of developing countries. Not surprisingly, China is one of the crucial players in the worldwide race to arrest global warming. The following section will highlight why not only China's high GHG emissions but also its weak environmental governance pose a serious challenge to any international regime to combat climate change.

China as a challenge to any international climate regime

Energy and GHG emission trends in China in worldwide comparison

In the inglorious ranking list of the world's top greenhouse gas emitters, China is catching up rapidly. China had overtaken the US by 14 percent in 2007 in terms of CO_2 emissions (Netherlands Environmental Assessment Agency, 2008) and makes up 22.7 percent of the world's total CO_2 emissions (World Resources Institute, 2010). But the picture looks different if one takes a closer look at China's per capita emissions, which are still below the EU 27 average (see Table 1.1), although they are expected to grow to 6.9 tons by 2035 (IEA, 2010b: 96).

As in most developing countries, the increase of GHG emissions is directly linked to economic performance. As we expect China's economy to grow considerably within the next decades, we expect the same for its GHG emissions. China's real gross domestic product (GDP) has grown by 9 percent annually since 1990, reaching US$6,282 in 2008 (IEA, 2010b: 602). Targets in the Chinese government's Eleventh Five-Year Plan (2006–2010) aim to increase GDP four-fold between 2000 and 2020 and to achieve average annual growth of 7.5 percent until 2010 (IEA, 2007: 284). In line with economic development, China's CO_2 emissions grew at an annual average of 5.6 percent between 1990 and 2005 (IEA, 2007: 313). The main source of CO_2 emissions in China is the power sector, which was responsible for 49 percent of China's CO_2 emissions in 2005. Its share is expected to grow to 52 percent in 2015 and to 54 percent in 2030 (IEA, 2007: 314).

Table 1.1 Total and per capita CO_2e emissions of China in worldwide comparison in 2007

Total CO_2 emissions (in Mt CO_2e and % of world's total)		Per-capita CO_2e emissions (in tons)	
China	6,702.6 (22.7%)	US	19.3
US	5,826.7 (19.7%)	Russian Federation	11.4
EU 27	4,064.5 (13.8%)	EU 27	8.2
India	1,410.4 (4.8%)	China	5.1
Russian Federation	1,626.3 (5.5%)	India	1.3

Source: Based on (World Resources Institute, 2010).

In line with its economic growth, China's electricity demand is expected to treble between 2008 and 2035, and would then require capacity additions equivalent to 1.5 times the current installed capacity of the US (IEA, 2010b: 98). If power generation continues to be heavily coal-dependent – a likely scenario giving China's vast coal reserves – consequences for GHG emissions and global warming will be alarming, as the share of coal in electricity production can be expected to increase in absolute as well as relative terms.

Scenarios for China's future GHG emissions offer an alarming picture: the latest projection of the International Energy Agency (IEA, 2010b: 96) expects that 58 percent of the world's CO_2 emissions in 2035 will come from China alone. With China's emissions increasing by 51 percent to 10.1 Gt in 2035, China's emission levels will exceed those of the OECD as a whole. Although China's per capita emissions will remain below the OECD average for the mid-term future, they are expected to grow to 6.9 tons in 2035 (IEA, 2010b: 96). As these scenarios of growing GHG emissions reveal, China is and will remain one of the most crucial arenas in the global battle against climate change. Only if emissions from China can be substantially reduced in the future can global warming be stabilized around the envisaged increase of 2°C.

Climate change as a challenge to China

Climate change represents a significant challenge to China itself – in terms of its negative local impacts, in terms of its position in the international climate negotiations, and in terms of its implementation of national policy targets on the ground.

Impacts of climate change on China

China has been assessing the impacts of climate change, its vulnerability, and its adaptation needs since the 1990s. Chinese studies generally acknowledge that climate change will have severe effects on China. China's National Climate Change Program lists the following expected effects of climate change (NDRC, 2007: 18ff).

- *Rising sea levels.* Over 18,000 km of coastline, including large coastal metropolises like Shanghai, and more than 5,000 islands are at risk in the event of a rise in sea level.
- *Increase in the frequency of extreme weather events and natural disasters.* This includes glacial retreat in north-western China, seasonal extremes of temperature, and unevenly distributed precipitation.

- *Desertification.* More than one-quarter of China's landmass is already affected by desertification. The increase in seasonal extremes of temperature and uneven precipitation will aggravate desertification and spread it to so-far unaffected areas.
- *Water scarcity.* In the next 50 to 100 years, the mean annual runoff is likely to decrease in several already arid northern provinces, such as the Ningxia Autonomous Region and Gansu Province. Precipitation is expected to increase steeply in a few already water-abundant southern provinces, such as the Hubei and Hunan provinces, leading to heavy floods.

International studies on climate change impacts also feed into the Chinese discussion. For example, the results of the report of Working Group II of the International Panel on Climate Change (IPCC) placed several Chinese regions among those most vulnerable to climate change, due among other reasons to a projected decreasing rate of precipitation and the negative consequences for agriculture yields (IPCC, 2001).

Chinese discourse on climate change

Climate change used to be a taboo subject in Chinese public discourse, because it was seen as part of a rising western eco-imperialism. Although the topic of climate change had already entered Chinese discourse by the late 1980s, when the first Chinese National Climate Committee was founded to coordinate climate change–related research, it was of concern only to a fraction of the nation's policy-makers (for example, to representatives of the State Science and Technology Commission and the National Environmental Protection Agency). At the local level, climate change–related measures were seen as barriers to economic development, and therefore were not in the interests of local political leaders (Pan, 2003). The interest constellation in China tends to focus on local economic development, because economic growth rates have for a long time been the central measure of the performance of local political officers, often determining decisions on promotions. Despite the discussion of introducing a 'Green GDP' (Oberheitmann, 2005: 16f) and integrating environmental performance indicators into the official promotion system (McGregor, 2006), the attitude towards environmental concerns has not changed much among local political elites. But there are several exceptions that show that green businesses linked with environmental protection can be consistent with local interests – for example, the decision of Baoding City, Hebei Province, to turn itself into a low-carbon city relying on the production of renewable energy equipment as a driver for its economic development (Qi et al., 2008: 13).

There was a fundamental shift in rhetoric when the issue of climate change became official state policy in 2005 (Jia, 2006). Still, the issue of climate change does not yet have the status of a stand-alone issue in politics, as it has in most OECD countries. Environmental awareness and knowledge on global warming and the consequences of climate change for China are limited in the general public, and thus climate change is not framed as an environmental issue in China. Even climate change campaigns of Chinese non-governmental organizations (NGOs) tend to link the issue to energy saving rather than global climate change (Schröder, 2008). Taking newspaper coverage of the issue as an indicator, the framing of the issue was one of 'foreign origin without much relevance for China.' The findings of a survey conducted in 2005 of the coverage of climate change–related topics in four newspapers – the *People's Daily*, the *Science and Technology Daily*, *Science Times*, and *Beijing News* – reveal that

> we find the Chinese media reporting of climate change has described the phenomenon as something certain but remote. It seems to be more of others' business so that neither Chinese scientists nor the public need to be involved. And if they do so, it appears, their involvement is to help defend the challenge brought by foreign competitors to curb the Chinese economy. (Jia, 2006: 3)

This conclusion is based on the following patterns (noted in Jia, 2006):

- All articles accept global warming as an accomplished fact.
- Few articles mention the impacts of climate change in China.
- Few articles analyze the role of China or other developing countries in global warming. Only six of 164 analyzed articles mention China to be the world's second largest CO_2 emitter.
- Most articles describe the Kyoto Protocol as a challenge to the Chinese economy.
- Some articles identify the CDM as a business opportunity.

The issue of climate change is thus rising on the political agenda, but it is not seen as a national priority. But there are some linkages to issues of greater immediate importance to China.

Air pollution is one of the major environmental problems China faces today. Only one-third of Chinese cities achieve an air-quality standard that does not represent a threat in the long run to the health of its inhabitants. Already in the mid-1990s, World Bank calculations estimated an annual economic loss of €6 billion as a consequence of urban air pollution, indoor air pollution, lead exposure, and acid rain (World Bank, 1997a: 23). A decade later, local air pollution is estimated to cost China in the range of 3 to 7 percent of its GDP each year (OECD, 2007b). The reasons for the bad air quality and the increase in acid rain are China's inefficient production plants and its dependence on coal, as coal burning is the main source of ambient SO_2, NO_x, and soot. This heavy local air pollution can be considered the main reason for the Chinese government's adoption of policies which – as a positive side-effect – also contribute to reducing GHGs.

China's fragile energy security is another reason why the Chinese government is interested in reducing its dependence on fossil fuels and increasing the share of renewable energies it uses. One negative side-effect of high economic growth rates has been a mismatch between demand and supply of power, leading to blackouts in 2003 and 2004 (IEA, 2007: 266). Blackouts were caused partly by electricity undersupply, and partly by lack of energy transport capacities. The latter represents a bottleneck, because the main industry agglomerations in China can be found along the eastern coast, while the main energy resources can be found in the interior. As a countermeasure, China plans the world's highest level of investment for the expansion of its energy infrastructure. In 2006 alone, 105 GW of new power plants (mainly coal-fired) were constructed (IEA, 2007: 266). Although most investments are earmarked for coal-fired power plants, China invested US$7 billion in renewable energies in 2006, representing 20 percent of the world's total. Renewable energies in China thus grew at an annual rate of 25 percent in the last years (UNIDO, 2007: 45). China's energy situation poses a heavy risk but also a great opportunity to the world. This dilemma is nicely summarized by Chris Flavin, Worldwatch president and lead author of 2008's edition of *State of the World*: 'On the one hand, China is close to passing the US as the world's biggest producer of carbon dioxide, and at the same time it is becoming an innovator in the field of renewable energy' (AFP News Agency, 2008).

Thus, in the Chinese discussion, mitigating GHGs is considered part of a larger endeavor to limit emissions of substances that lead to local air pollution, to improve energy efficiency and thus save on energy costs and contribute to China's long-term energy security, and to set up a domestic renewable-energy industry and transform it into another market in which China will take the lead (Jacob et al., 2005) – instead of leaving the floor to foreigners, as happened with the Chinese automobile industry (Chin, 2010). Although political campaigns of the Chinese government in these three areas are not initiated primarily for reasons of climate protection, they have positive side-effects for reducing China's GHG emissions.

China's position in the international climate change negotiations

In the international climate negotiations, China's diplomatic stance mirrors that of most developing countries. When the issue of climate change entered the international political agenda, the Chinese government was skeptical about this issue, raised by industrialized countries, and suspected and condemned attempts to 'intervene into sovereign affairs' – for example at the UN Conference on the Human Environment in 1972 (Bechert, 1995; Ross, 1998: 137). In the 1990s the Chinese position became cooperative, and China has been among the first countries to sign the United Nations Framework Convention on Climate Change (UNFCCC) and the Kyoto Protocol.

China's position towards the CDM was also skeptical at the beginning, as it was regarded as another attempt by industrialized countries to use the resources of developing countries to forfeit their obligation to take up mitigation measures at home. Hoping for technology transfer and additional foreign investment, China eventually viewed the CDM favorably and – with the support of foreign donor programs – set up the necessary institutional structures and regulations for becoming a full-fledged host country to the CDM.

In the debate over a post-2012 climate regime, China has increasingly come under pressure to commit itself – along with other large GHG-emitting developing countries – to binding reduction targets. Pressure mainly took the form of addressing the topic of climate change during state visits. For example, EU Commission President Barroso, Germany's chancellor Merkel, and France's president Sarkozy all addressed climate change during their state visits to China in 2007. This pressurizing strategy is allegedly paying off, as one interviewee explained:

> Every time foreign leaders come to China, they want to discuss climate change. Every time Hu Jintao goes abroad, foreign leaders want to discuss climate change. It is really getting on his nerves. The Chinese delegation submitted this very constructive proposal to the Bali COP/MOP, because the leader had said 'you must do something to reduce the pressure on my shoulders'. (Interview 37)

Being part of the G77 and referring to the principle of common but differentiated responsibilities, the Chinese government has so far denied any international reduction commitments and opposes any binding international targets and timetables, but has pledged to reduce GHG emissions through ambitious national policies. Without large GHG reduction commitments from the US, experts do not expect the uptake of binding targets by China (Zhang, 2008: 4ff). China is nevertheless trying to be seen as a constructive player in the post-2012 negotiations: it has softened its rhetoric and has indicated its readiness to take on sectoral commitments in return for technology transfer (Höhne et al., 2008: 83). Together with the G77, China is, for example, demanding the establishment of new and additional multilateral funds to foster technology transfer and the establishment of a technology transfer mechanism (G77 & China, 2008; Watanabe et al., 2008: 12). China also spares no opportunity to stress its own national efforts; for example, at the COP16 in Cancún, Mexico, Xie Zhenhua, vice-chairman of the NDRC, explained:

> The Chinese Government has identified proactive tackling of climate change as a key strategy for economic and social development. It has formulated the National Plan on Addressing Climate Change, and put forward a target in the Eleventh Five-Year Plan period to reduce the energy consumption per unit of GDP by 20 percent in 2010 [from] the level of 2005. Notable headway has been made after adopting multiple policy measures and action, which include optimizing industrial structure, eliminating outmoded production capacity, enhancing energy conservation and improving energy efficiency, accelerating development of clean energy and increasing forest carbon sink, etc. ... That is a great contribution to global efforts in tackling climate change and will stand the test of history. (Xie, 2010)

When analyzing the Chinese position in the international climate negotiations, one has to be aware that, although the Chinese delegation speaks with one voice during the official discussions, within China's government different perspectives compete. Because of the potential goal conflict between climate protection and economic growth, the Chinese government favors 'no regret' options that contribute to climate protection without harming the economy. As long as measures for climate protection have positive effects on economic development, or at least have no damaging consequences, the Chinese government – like most other governments – is keen to position itself as a climate protector. As a mechanism that can bring foreign investment into China in return for emission reductions, the CDM could become an example of the compatibility of environmental protection with economic growth. If the CDM experience turns out to be positive, chances are good that important countries like China will adopt a more proactive stance in the post-2012 climate negotiations.

China's policies for reducing greenhouse gas emissions at the national level

While there has not been much change in the official Chinese position in international negotiations for the post-2012 Kyoto Protocol, China is setting up several targets at a national level that contribute to climate protection. Chinese Five-Year Plans tend to acquire a green edge over time: the Ninth Five-Year Plan (1996–2000) started with a demand for a decoupling of economic growth from environmental degradation; the Tenth Five-Year Plan (2001–05) included a target to reduce SO_2 emissions by 10 percent in 2005 relative to 2000 levels; the Eleventh Five-Year Plan (2006–10) envisages a 20 percent reduction in the energy intensity of production and in polluting emissions by 2010 relative to 2005 levels; and the Twelfth Five-Year Plan (2011–15) even includes a target for a reduction in carbon intensity.

The central policy related to climate change, but based on energy security considerations, is the 'Save Energy and Reduce Emissions' (*jieneng jianpai*) campaign. The Comprehensive Action Plan for Energy Saving and Emissions Reduction was announced on September 1, 2007 by the NDRC and 16 other ministries, and foresees energy-saving and emission-reduction targets for nine areas – for example, by introducing energy codes for buildings and the labeling of household equipment, and by enforcing energy-efficiency regulations for public procurement. According to an estimation of the US-based NGO Energy Foundation, the campaign could reduce up to 1.2 gigatons of CO_2 emissions from the baseline (Qi et al., 2008: 5). One central measure of the campaign is the 'Top 1,000 Enterprises Energy Efficiency Program,' under which contracts and targets for more energy efficiency have been drawn up with the 1,000 largest companies in the country. However, the program seems to have faced some implementation difficulties, as expertise on how to implement energy-saving measures in the industry still needs to be developed (IEA, 2007: 276).

The campaign's targets have been allocated among provinces and industrial sectors by the NDRC, so that annual figures for energy consumption per unit of output have to be made public for all regions and major industries, and energy efficiency improvement is now among the criteria used to evaluate the job performance of local officials. While the overall reduction target for energy intensity in the whole of China is 20 percent, the targets vary widely between the provinces: 12 percent for Tibet, 17 percent for the remote western provinces of Sichuan and Qinghai, and up to 30 percent for the economically advanced eastern province of Jilin (IEA, 2007: 409). Preliminary data reveals, however, that the 20 percent reduction target has been missed, although only slightly. Between 2005 and 2009 energy intensity fell by 15.6 percent, though it increased slightly in early 2010 (NBSC, 2010). While the national target was almost met, most areas have further to go before meeting their provincial targets. In only two provinces – Beijing and Fujian – did the decline in intensity in that year meet or exceed the target (IEA, 2007: 410). Among measures such as the closure of small coal-fired power plants and inefficient industries, the CDM is regarded as one instrument to attract investment in clean energy infrastructure.

This illustrates some typical features of environmental policy-making in China: national targets are decided by leading decision-makers at a central level, and given out as 'directives.' These targets are then broken down into provincial-level targets, and the responsibility for their implementation is passed to the provincial political leaders. These, in turn, pass energy-saving and pollution-reduction targets to the prefecture and county level, and also to state-owned enterprises (SOEs) (Qi et al., 2008: 6). This combination of central command with contrasting local interests is often not very conducive to effective policy implementation.

Another national set of policies directly related to climate change is contained in China's National Climate Change Program, which specifies the following climate-related targets: the reduction of energy intensity and the promotion of various renewable energy sources such as hydropower and bio-energy, the development of highly efficient thermal power technology, and the promotion of nuclear power. An institutional change that reveals the upgraded status of climate protection is the formation in June 2007 of the National Leading Group on Climate Change, with Premier Wen Jiabao as its director. Although on paper these targets are ambitious, doubts remain as to whether they are realistic. Can we indeed expect their timely and effective implementation?

Climate change as a challenge to China's system of environmental governance

Although China's performance is spectacular in terms of economic growth and the improvement of general living standards, one area in which it constantly underperforms is the environment. China has become infamous as one of the world's most polluted places, and most scenarios show deteriorating trends (World Bank, 1997a; Smil, 2003; Economy, 2004; Turner and Linden, 2007; IEA, 2010b). Why is the Chinese governance model unable to deliver the increasingly valuable public good of an intact environment? What are the reasons for its weak performance? If China has done so poorly with regard to the environment in the past, what can we expect for its efforts to contribute to climate protection? Fortunately, there are many signs that China's system of environmental governance is in transition at the subnational, national, and international levels (Mol and Carter, 2007: 3). When we analyze the patterns of state and non-state climate governance in China, we have to be aware that we are dealing with a moving target, because the Chinese economic system is still in transition (Chu, 2010). China's approach to economic transition is unique in attempting to combine a centralized political system with an increasingly decentralized, market-oriented economy. The transition towards a market economy requires a redefinition of the role of the state. But in which direction are these transitions heading? China experienced several different phases of political, fiscal, and administrative decentralization, and went through several phases of privatization, ownership diversification, and corporatization. The official goal of the transition process is not that of becoming a capitalist country; rather, the government aims at 'socialism with Chinese characteristics.' This concept has its roots in the work of Deng Xiaoping (Deng, 1983 and 1987), who initiated several decentralization and privatization measures in order to provide incentives for local government to foster local economic development. Is the turn towards a market economy and the introduction of market mechanisms a chance to improve China's environmental governance? Is the turn towards the market a cause for optimism that China is becoming not only willing but also able to enforce climate-related policies?

This question can be addressed by first providing an overview of China's traditional system of environmental governance, and then describing which features of change can be observed. Special focus is placed on experiments using market mechanisms to achieve environmental targets in China's economy in transition, and to describe the intricate and changing relations between public and private actors in China.

Development priorities

The Chinese governance system has traditionally been very state-centric. The central government has usually defined targets, and implementation was left to local-level administrations. In the past, the main reasons for the failure to achieve adequate environmental governance were the targets themselves. Thus, it can be said that the lack of political will to pursue environmental protection was, until the 1990s, the most important factor causing ineffective environmental governance. Political priorities were set on economic growth, and the unsustainable use of natural resources was seen as an acceptable side-effect (Lothspeich and Chen, 1997; Shapiro, 2001). Although the objective of economic growth has become less ideological and more quantitative (for example, the Eleventh Five-Year-Plan [2006–10] specifies a target of 7.5 percent annual GDP growth), the achievement of economic growth is still regarded as the main pillar of legitimacy for the Communist party government. But, within the last decade, other policy objectives have gained prominence on the political agenda in China. The focus on economic growth came to be contested with the deteriorating state of the environment in China, a growing global scarcity of energy and natural resources, and the resulting energy insecurity and environmental calamity for China (McElroy et al., 1998). Since the 1990s we can thus detect at least a rhetorical shift in favor of environmental protection. China's Five-Year Plans and an upsurge of environmental regulation reflect the rise of the issue of environmental protection on the political agenda. Over time the issues of environmental and climate protection have – at least rhetorically – gained equal status with the traditional government goal of economic growth (Zhang et al., 1999). The equal standing of the two objectives of environmental protection and economic growth has been restated by Chinese leaders – for example by Wen Jiabao, premier of China since 2003, at the Sixth National Environmental Conference in 2006 (Lan et al., 2006: 8).

Recognizing that economic growth measured purely in terms of the rise of GDP does not necessarily mean an improvement in welfare for the general population, the Chinese government has embraced the 'scientific development concept,' which aims to 'shift China's economic growth from a resources-based pattern to an innovation-driven pattern and to promote the notion of scientific development in both economic and social areas' (Wang, 2009: 9). This concept promotes a more qualified economic growth model that incorporates sustainable development, a harmonious society, and the achievement by 2020 of a 'Xiaokang Society' – a society in which most people are moderately well-off and a firm middle class is established. This new outlook has resulted in an economic transformation away from a centrally planned towards a market-oriented economy, and is accompanied by decentralization measures, bureaucratic reorganization processes, and a thoroughgoing opening up to and integration in the world economy (Mol and Carter, 2007: 3; Breslin, 2009).

Inadequate policy implementation

China's history of environmental protection is unfortunately a story of mixed results. There are positive examples of what has been achieved – for example, the world's highest reforestation rate. But the number of failures in environmental governance – despite allegedly good intentions – is devastating, and calls for a restructuring of China's environmental governance system. Lack of effective implementation seems to have been the main reason, as it continues to be, for the deteriorating environmental situation in China, which has left it on the edge of environmental catastrophe. In the past, command-and-control instruments have often failed completely.

One typical example of the Chinese struggle to fight environmental degradation with command-and-control instruments is the decade-long attempt to take control of SO_2 emissions, which cause acid rain. China's Ninth Five-Year Plan (1996–2000) set the national target for SO_2 emissions to come down to 23.7 million tons by 2000 (Yang and Schreifels, 2003: 10). This national target was in turn allocated to China's 31 regions and provinces, which in turn then distributed emission-reduction targets among their major industries. A levy system was introduced, fining industries that did not comply with their target. Although the national SO_2 emission charges generated a total of €92 million within a decade, the system did not provide economic incentives to curb SO_2 emissions, because the emission charge of €0.16 per kilogram of SO_2 is less than the average marginal abatement cost for SO_2 emissions (Yang and Schreifels, 2003: 9). Most companies therefore preferred to pay the fine instead of curbing their emissions (Sinkule and Ortolano, 1995), or the levies were not collected at all due to the bureaucratic weakness of the responsible Environmental Protection Bureaus (EPBs) (Tang et al., 1997). The emission-reduction target was therefore not reached: in 2005, SO_2 emissions reached 25.49 million tons, representing an increase of 27 percent since 2000.

The strong role of the central state

Compared to other countries with economies in transition, China is still a very 'statist' society, although state actors now have increasingly to operate within a market framework (Lane and Myant, 2007: 2). Embedded in a tradition of 3,500 years of rule by a state bureaucracy,

China's form of state governance is still highly technocratic. This also applies to the environmental field, in which the Chinese governance system has been described as 'state-driven environmental management' (Morton, 2005: 3). The government's centralized mode of regulatory control is still dominant in environmental governance, although new forms of governance are emerging that incorporate market forces and public participation. China also traditionally pursues a technocentric approach to environmental governance, believing in its ability to solve environmental problems through improved technologies and engineering capabilities (Morton, 2005: 5). Such a top-down approach to environmental governance has also fostered a popular belief that environmental protection is essentially a government responsibility (Morton, 2005: 5). Attempts to use the government's state propaganda machine also for environmental education (all over China, posters advertise the national 'energy saving and emissions reduction' policy) have strengthened such perceptions.

The undersupply of public good 'environment'

Despite its strong state, effective solving of problems and achievement of targets within environmental governance is weak in China. In its traditional, state-focused environmental governance system, China has experienced inefficiencies and ineffectiveness comparable to the 'state failure' that European countries were struggling with in the 1980s (Jänicke, 1990). The state is not sufficiently able to provide for the public good of an intact environment. The environment can be counted as a public good because of its characteristics of non-rivalry and non-excludability in consumption (Ostrom, 1990; Cornes and Sandler, 1996). The provision of the public good of the 'environment' is generally regarded to be a core task of governance and a responsibility of the state (Risse and Lehmkuhl, 2006: 15). In China, the political will to achieve better environmental governance can, at least to an extent, be taken as a given. As we have seen, policies in China have acquired a 'green edge' over time. Moreover, the Chinese delegation in the climate negotiations has boasted that China is already doing well in addressing climate issues in its national policies. While this is true for policy design (although there are still many problems with insufficient planning capacities and a lack of policy coherence and coordination – see Lampton, 1987: 14 on fragmented authoritarianism in China), policy implementation unfortunately looks very weak on the ground. China's weakness in policy implementation is one fundamental cause of its governance failures in environmental protection. What are the problems with environmental governance in China, and how might they be overcome? The following section gives an overview of the enforcement problem and how it might be overcome through the increased use of market mechanisms and the rise of the private sector as an alternative to the inefficient state sector.

Problems of rule-enforcement

Though topics like energy efficiency, clean power generation, and climate change have become a substantial part of Chinese political rhetoric, experts warn against inflated expectations for their achievement, and identify a growing gap between ambitious rhetoric and ineffective policy implementation on the ground. Policy implementation is a well-known challenge in Chinese environmental governance (Lieberthal, 1997; Economy, 2004 and 2007; Turner and Linden, 2007; Schreurs, 2008: 351). According to Wang Canfa, an environmental lawyer, barely 10 percent of China's environmental laws and regulations are actually enforced (Economy, 2007). The building of institutions – and, in particular, their effective operation – is a challenge that becomes even more daunting with the increased use of market-based environmental policies, for which functioning institutions are an enabling condition.

Insufficient local state capacity

There are many causes of problems with enforcement, of which insufficient state capacity, or even state failure, seems to be the most crucial. State capacity can be defined as the ability to 'undertake collective action at least cost to society' (Cheung and Scott, 2003: 6), or 'the ability of a government [to get] its job done,' and includes its capacity to mobilize society, extract resources, steer development, and legitimize the regime (Wang and Hu, 1993: 235). In China, state capacity in the enforcement of environmental regulations is considered to be weak, especially at the local level (Chan et al., 1993; Wang, 1995; Wang and Hu, 2001; Jia and Lin, 1994; Schwartz, 2000).

One fundamental problem of environmental enforcement in China is that targets are set centrally, while concrete measures for financing and meeting central targets are left to local administrations, which are often neither willing nor able to ensure enforcement. A lack of will is caused by missing incentives for local government to enforce rules, whose priority of economic development often ties it very closely to the industries and companies on which it would have to impose fines for environmental pollution. Consequently, if policy priorities of the different levels of government do not match, deadlock is the result

(Sinkule and Ortolano, 1995: 10). Government departments also compete horizontally both for resources and status. For example, the work of EPBs is often stalemated, because they are instructed by the national Ministry of Environmental Protection but receive their resources from their province- or county-level government, which usually regards economic growth as its priority and allocates financing accordingly (Walder, 1992; Tang et al., 1997). Local governmental officials are also torn between, on the one hand, their interest in supporting clean industries and renewable energy, which can provide a 'green' government image and generate lucrative sources of tax revenue, and, on the other, their interest in keeping old and polluting power plants running so as not to to impede the economy, which might negatively influence their performance evaluation (UNIDO, 2007: 46). But even when motivation is high, government capacity to implement policies is often insufficient due to insufficient human, technical, and financial resources, and to a lack of coordination between state agencies, which instead often compete openly for influence and revenues (Sinkule and Ortolano, 1995; Jahiel, 1997 and 1998; Ma and Ortolano, 2000).

Public sector reform

There are three basic reform approaches aimed at overcoming insufficient local state capacity that are recommended for China, but also for other East Asian states suffering from local governance failure (Cheung and Scott, 2003: 3). These are, first, the strengthening of state capacity; second, new public management–style reforms strengthening reliance on the market; and third, the involvement of civil society in governance. These recommendations have so far produced only limited results in the improvement of environmental governance in China. One can observe many efforts to strengthen state capacity, some experiments with engaging the market, and rare measures to allow more civil society participation. Of course, these three reform approaches cannot be handled one by one, but must be integrated in a comprehensive manner and adapted to local circumstances. The following paragraphs will focus on how reform of the public sector and increased involvement of the business community through market mechanisms like the CDM can improve rule-enforcement.

The first reform approach is to strengthen state capacity in order to enforce environmental regulation, and to oversee the implementation of policies and programs effectively and independently of special interests. It also entails capacity development for establishing transparent, accountable, predictable, participatory, efficient, and corruption-free government institutions. The ultimate aim of this approach is to achieve good governance (Pierre and Peters, 2000: 65). The reform of the public sector in China entails both a strengthening of state capacity and a streamlining of the public sector by outsourcing tasks to the private sector. One major reform measure to improve state capacity has been the introduction of managerial tools from the private sector into the public sector, replacing the traditional administrative approach. This includes the introduction of economic leverage and market instruments, 'load-shedding' (*tiao baofu*), and privatization (Lee and Lo, 2001: viii). Although many of the reforms intended to increase state capacity resemble routine new public management practices, the focus of the reforms is still more on strengthening bureaucratic performance than on outsourcing state tasks to the private sector. Many efforts have also been made to provide incentives for governmental officials to implement policies. The cadre performance evaluation system has been reformed to include a responsibility for the enforcement of environmental regulations (Chou, 2004). Despite these efforts, reforms still have a long way to go before the goals of bureaucratic accountability and the eradication of corruption are achieved. 'Governance with Chinese characteristics' entails an enhanced role for the state (Burns, 2003b: 68). Consequently, a widespread debate has emerged among Chinese experts and researchers on how to redefine the role of the state in a market economy (Burns, 2003a; Yang, 2004; Burns, 2003b), especially concerning the role of the state in local governance (Wong, 1987; Zhong, 2003; Caulfield, 2006; Chen et al., 2002).

The reform of public management in China was launched in the early 1980s with an extensive privatization of many tasks, such as welfare and housing, and of many entities, including the former SOEs (Lee and Lo, 2001: viii). In order to reduce administrative costs and to concentrate on core tasks, many tasks have been assigned to allegedly non-state actors. The challenging dimension of this reform, however, is that most of these actors remain somehow linked to the government. For example, while the number of administrative staff on the public payroll has declined considerably in the last decade, the staffing of the semi-public *shiye danwei* (see below) have increased dramatically. The transfer of employees between the formal and informal public sectors has been one option open to provincial government officials striving to adhere to national requirements for downsizing administration while simultaneously avoiding a rise in unemployment and discontent in their jurisdiction (Lo et al., 2001b).

The rise of the private sector

The second reform approach aims to mobilize market forces for the provision of environmental services and public goods. This might include the provision of economic incentives for environmentally friendly behavior or the outsourcing of environmental service provision to private companies. The role of the private sector has experienced a 180-degree turn: while it was most despised in communist times under Mao Zedong, it was rehabilitated under Deng Xiaoping's reform movement, and has become the engine of growth in today's China.

Until the opening of China's economy to the outside world in 1978, most of its means of production were state-owned. The rise of the private sector started with Deng Xiaoping's dictum that 'it doesn't matter whether the cat is white or black as long as it catches mice' on his South China tour of 1992. The 15th Congress of the Chinese Communist Party, in 1997, further reduced legal and economic barriers to private ownership (Lau, 1999). In order to draw in foreign investment, the Chinese government gradually introduced private property rights for foreign companies, and, in 1998, finally for Chinese companies and citizens as well. In a manner reminiscent of both communist and Confucian ideology, private companies were at the bottom of the hierarchy of market actors, while SOEs enjoyed preferential status, with easy access to state capital (Huang, 2002). One crucial question in the Chinese transition towards a market economy is that of how to deal with the often loss-making SOEs. The first step in SOE reform was to introduce profit-based incentives; for example, companies were allowed to keep profits from production that exceeded the centrally endorsed quota. A second step has been a change in ownership structures: SOEs were allowed to be turned into stock companies with initial shares divided by employees, the local government, and individual investors. Up the time of writing, collective enterprises still exist as township-and-village enterprises (TVEs), which show considerable profitability. The restoration of private business was finally accomplished by private business owners being allowed to become members of the Communist party, which enhanced their social and legal standing (Tsui et al., 2006: 6). The business sector is also envisaged as playing a more active role in environmental management, – for example, by observing and disseminating good practices in the area of environmentally friendly production methods.

Although the Chinese economy has seen major changes in property rights and the status of private companies, the country is still far from a full-fledged market economy (Powell, 2008: 15). In stark contrast to Hong Kong, which consistently takes first place, the PRC only ranked

140th out of 179 countries in the 2010 economic freedom ranking published by the free-market Heritage Foundation (Heritage Foundation, 2010). This 'economic freedom' score has, however, improved considerably since the initiation of China's policy of opening up to foreign trade, and especially since China's entry into the World Trade Organization (WTO) in 2001. Within China, market conditions differ radically between regions: those with the highest degree of 'marketization' are the areas where most of the economic growth has occurred. The private sector has become China's growth engine, accounting for 59.2 percent of China's gross domestic product in 2003 (OECD, 2005c: 125). On the other hand, the share of industrial output accounted for by SOEs dropped from almost 100 percent at the start of economic reform in 1978 to 33 percent in 2003. While SOEs still contributed 50 percent of China's GDP in 1990, their share had fallen to roughly 30 percent by 2008 (Tsui et al., 2006: 5; NBSC, 2009).

Private and public relations in flux

The coexistence of various ownership systems is typical of the Chinese economy, because China takes a two-track approach towards economic transition (Lai, 2006). New policies are tried out in one sector in the context of pre-existing regulations, leaving it to the actors involved to decide which track to follow. This approach enables the introduction of controversial reforms because it does not create losers within the bureaucracy (Tenev et al., 2002: 9). In line with this 'dual-track' approach, the Chinese government supported private ownership while most of its industrial output was still achieved by the SOEs. Support was also given to cooperative arrangements between public and private entities. In 1994, for example, the government launched the 'Brilliant Cause Program,' which encouraged private enterprises to implement projects within public-private partnerships (Wang, 2001: 93).

This coexistence of public, private, and hybrid forms is typical for economies in transition, but makes categorizing actors according to their status a difficult task (Chu, 2010). The private sector can include private firms, household businesses, and township-and-village enterprises, but also the privatized SOEs and their spinoffs. The latter two categories, in particular, are hard to classify, because both state entities and private companies have incentives to disguise their status. State entities have private spinoffs, because these enable them to make profit and earn money. Private companies often claim to have government affiliation – they wear 'red hats,' thus 'donning the garb of correct political status' (Tsui et al., 2006: 6) – because it gives them more political clout and better standing, and provides soft assets such as information and networks. Private firms thus often attach themselves to state-owned firms in order to avoid discrimination or to gain privileges. Others register as township-and-village enterprises to gain protection or seek support from local government (Tsui et al., 2006: 6, Nee, 1992). Due to these practices, the real status of the private sector in China is hard to pin down, because the fluidity of firms' status and ownership category makes tracking them over time very difficult (Tsui et al., 2006: 5). Ownership forms in China's transition period resemble amphibian forms between state and private ownership (Ding, 1994); or, as Nee has nicely phrased it:

> The transition economy has given birth to a new diversity in organizational forms and a plurality of property rights. The spectrum spans the continuum from the formal and hierarchical state-owned enterprises to small family-owned firms run by peasant entrepreneurs. (Nee, 1992: 2)

Shiye Danwei: public service agencies

The 'Shiye Danwei', a very Chinese kind of agency, is the most prominent form in which the hybridization of state and private ownership and control has become institutionalized. Because many of the provincial CDM centers belong to this kind of unit, this organizational type will be briefly outlined. A Shiye Danwei is defined as a 'social service organization established by the state for the purpose of social public benefit' (State Council, 1998). Such an organization is run like a company, but implements public projects and services. Depending on the local context, it operates as the commercial branch of a government department, or it can come close to being a private company that is assigned jobs mainly by the government. There is a high degree of heterogeneity among public service units: some solely provide public services such as disease control, while others provide only private goods, such as books. The *danwei* ('unit') system is a relic of the Chinese Communist system, in which there were four types of *danwei*: industrial units (*qiye danwei*), agricultural units (*nongye danwei*), service units (*shiye danwei*), and administrative units (*xingzheng danwei*). In the past, these units received their orders from the central government and were de facto branches of the Beijing government (OECD, 2005a: 13). Nowadays, the unit system has been dissolved for the industrial and agricultural units, but the other two still exist. There are about 1.3 million public service units, employing more than 25 million people, which make up about one-third of public sector employees. The most common areas in which these units are working include education (school, universities) and health (primary care centers, hospitals), but they also include infrastructure and research (OECD, 2005b: 81). Lam and Perry claim that the service organizations 'find no parallel in Western countries, although many organizations falling into this category surely exist in Western countries (Lam and Perry, 2001:19).' Because the public service units are one of the last big relicts of the planned economy, in which the state alone was responsible for the provision of public goods, these units are a core focus of reforms towards a market economy. The Chinese government is thus in the midst of its process of decision-making on whether to retain ownership of public service delivery organizations, or whether to switch to employing private firms for that purpose. A decision on what role to assign to the public service units in the future will mark a turning point in the debate on the role of the state in the emerging Chinese market economy and on the modes of governance in China (OECD, 2005b: 81).

Market mechanisms as a remedy?

Promotion of market mechanisms for developing countries

Market-based instruments have long been fashionable in OECD countries as cost-efficient means to internalize the environmental costs of economic processes, thereby overcoming the old antagonism between economic development and environmental protection. Environmental protection is now widely seen as a necessary prerequisite for long-term economic development (Arrow et al., 1995). Not surprisingly, environmental economists have been able to establish theirs as the mainstream approach in the debate over how to reconcile these two fields of enquiry (Pearce et al., 1989; Tietenberg, 1990; Hanely et al., 1997; Stavins, 2000). The basic idea behind using market mechanisms for environmental governance is that they provide economic incentives that change polluters' behavior while maintaining flexibility as to how the desired environmental objectives are to be achieved (Shogren, 2007: 328).

Similar advantages of economic instruments can be expected for developing and newly industrializing countries. Since the 1990s, economists have also recommended market-based instruments for developing countries as means of achieving sustainable growth patterns (Panayotou, 1998; World Bank, 1996; World Bank, 1997b; Huber et al., 1998). The major development cooperation agencies also introduced market-based reforms through their aid programs, and often attached related conditionalities (Minogue, 2001: 34). Recommendations usually include the restructuring and reduction of the public sector, particularly through privatization; the reorganization and slimming down of central civil services; the introduction of competition into remaining public services; and efficiency improvements through performance management and auditing (Minogue, 2001: 21). Market-based instruments have also been promoted for environmental governance – for example, emissions trading has been recommended since the 1990s for developing countries and countries in transition as a suitable mechanism for tackling growing environmental pollution (Dudek et al., 1992; Potier, 1995; ADB, 2001).

But are the positive outcomes for market mechanisms in OECD countries so easily transferable to developing countries? To be applicable in developing countries, market-based instruments have to be adapted to the special framework conditions of countries that might not have a fully liberalized market, fully privatized companies, or a government with good capabilities in the setting and implementation of rules. From a more technical perspective, polluting companies and their supervising environmental authorities need to be able to monitor and measure pollution, to understand the relationship between production and pollution, and to translate the social costs created through resulting environmental damage into monetary terms, based on which political decisions can then be made (Larson and Bluffstone, 1997: 7). This perfect market situation is hardly a given in industrialized countries, but is even less prevalent in developing countries, where information on pollutants and their consequences is often imperfect, and pollution charges often do not provide sufficient incentives for pollution abatement (Larson and Bluffstone, 1997: 16f). The introduction of the CDM as a market mechanism in a diverse array of developing countries should therefore furnish an instructive example of how market mechanisms perform in the economies of developing countries.

Chinese experiments with market mechanisms

Since the millennium, China has increasingly experimented with market-based approaches, although still on a pilot level. The use of market mechanisms and the general transition towards a capitalist economy are not necessarily seen as contradicting the state ideology of 'socialism with Chinese characteristics,' due to the pragmatic approach welcoming all means that are effective in reaching policy objectives.

On the contrary, China pursues a mixture of command-and-control and market-oriented instruments in order to cope with the increasing damage to its environment. Economic instruments were seen as a means of overcoming the rigidity, complexity, and wastefulness of the regulatory and administrative methods (Lee, 1987). In the last three decades, China has tried several economic instruments for environmental governance: price regulations, subsidies, emission charges and standards, environmental funds, and taxes on natural-resource consumption (Economy, 2007). China began to experiment with economic instruments just four years after 1978, when it opened its economy to the outside world. In 1982, the Chinese government began to impose pollution charges, levied as non-compliance fees on both the quantity and the concentration of discharges into the atmosphere, as well as in water, noise, solid waste, and radioactive wastes (O'Connor, 1996: 10). While initially ambitious, the system turned out to be inefficient due to the low level of fines, the poor capacity of governmental officials to enforce the regulations, and the possibility for state-owned enterprises to hand down costs from the emission levies to consumers. Similar negative outcomes have resulted from charges on SO_2 emissions. Because the classical command-and-control instruments of standard-setting and pollution fines failed, China is now experimenting with economic instruments for SO_2 emission reductions at both national and the provincial levels (Yang and Schreifels, 2003). Instead of the old levy system, companies are now rewarded for their positive environmental measures by tax reductions and low-interest loans. But because China still has many SOEs whose business decisions are ultimately guided by governmental officials, the application of market instruments to public enterprises often leads to a paradoxical situation in which market-based instruments are not able to work effectively (Sicular, 1996).

While some experiments with market mechanisms have shown positive results at first, one unresolved problem is that of strong state intervention. Although there are some economists who argue that a strong state is necessary, especially in developing countries, to uphold the public interest (Stiglitz, 1989; Chandler, 1977), state intervention is usually seen as the last resort. A situation demanding state intervention is that of market failure, which refers to 'conditions under which a market economy fails to allocate resources efficiently' (World Bank, 1997b: 26). This can include situations of incomplete markets and imperfect or asymmetric information, which in turn lead to inefficient outcomes. If market failures and the existence of externalities lead to socially undesirable market outcomes, intervention in the market can even become an imperative (Jaffe et al., 2002: 48). While intervention as such is acceptable from a theoretical point of view in cases of market failure, it remains to be seen whether public agencies or private companies function as better market facilitators in such cases.

Provincial CDM centers as a policy experiment

These considerations make it clear that China's transition process is a huge national experiment. Because there is no blueprint available and no historical precedent for a 'socialism with Chinese characteristics,' Deng Xiaoping had already characterized the task of development as 'crossing the river by feeling for stones' (*mozhe shitou guo he*). As a consequence, reform initiatives tend to be incremental, adopting a step-by-step approach. Various policy packages might be integrated into a systematic whole through trial and error (Lee and Lo, 2001: 5). A typical feature of the Chinese economic transition process is its gradualism – its process of 'growing out of the plan,' which includes many small steps without any major revolution (Naughton, 1995). Thus, the Chinese transition from a planned towards a market economy can be seen as a flexible process incorporating evolutionary changes in institutions and policies (Walder, 1996: 2). Typical of the Chinese approach of incremental reform are local experiments with pilot policies (Goldstein, 1995; Shiu, 1997). For any policy problem, a promising policy design is first trialed at a local level – for example, in one province. Once the policy has been successful, it is gradually extended to other provinces. If it succeeds in other locations as well, it has a good chance of becoming a national regulation or law. Because Chinese economic administration is decentralized, there is fierce competition between local governments. This regional competition functions as one of the most important drivers of innovation and economic growth in China (Heilmann, 2008).

Local experiments are thus important in China for determining the future direction of national policy decisions. If a local policy experiment is successful, chances are good that it will be expanded and implemented on a national level (as suggested by Qi et al., 2008: 396 with regard to local climate policy innovations causing national-level policy change). Pilot projects with market mechanisms also reveal new developments in the changing relations between state and market in China. China has a legacy of state interventionism, despite being in transition towards a market economy. The CDM is an international market mechanism that has been, to put it bluntly, imposed on China, so it provides a crucial window onto how market mechanisms designed according to OECD experiences work in a country in transition. At the national level, a 'market capture by the state' might be one possible result, as the Chinese government has been apt to exploit the CDM for its own political purposes (Schroeder, 2009). It will be interesting to see whether such strong state interventionism is also reflected at the local level, and whether it does any good to the local CDM market. It is no surprise that the institutions in China that support CDM markets are not private enterprises, but usually have a strong government affiliation or are even parts of local government. The provincial CDM centers are an example how a local policy experiment succeeded and is now being implemented on a national scale in 27 provinces. Why has the Chinese government considered these CDM centers to be worthwhile for national-scale replication? Approaching them as a case study, what generalizations can be drawn from their structure and effectiveness about the performance of semi-public agencies in facilitating nascent markets?

A comparative case study investigating the effectiveness of provincial CDM centers as semi-public agencies in delivering public goods makes an interesting case for the more general discussion of whether state intervention is necessary or successful in facilitating markets. Since the CDM centers are in direct competition with private business actors, this analysis can also provide some insights into the question of whether public, private, or even hybrid actors are better able to provide public goods for nascent markets. The public good in this case is the provision of information and capacity development, which would support CDM diffusion at the local level. CDM centers can thus be seen as institutions intended to reduce transaction costs for market participants by providing these public goods. Ultimately, the CDM centers' contribution to CDM market development will enable CDM projects that reduce GHG emissions, and thereby contribute in principle to sustainable development and environmental protection.

The following chapters will show that, because provincial CDM centers take a variety of intermediate shapes, from purely public to almost pure private forms of organization, they are a good example of the 'organizational zoo' that has resulted from the indecision on how to move forward with the delivery of public services in China. In a way, the case studies on how local CDM markets and their institutions have been created echo the challenge China faces in creating new institutions for a market economy.

Part II

The Performance of CDM Centers as Semi-Public Agencies

2
The Need for Capacity Development in the Early CDM Market

China's CDM market

This chapter gives an overview of the CDM market in China and outlines the puzzle of the varying effectiveness of provincial CDM centers in CDM diffusion and market facilitation. With a focus on the provincial level, the market development between 2005 and 2010 is described, market barriers are identified, and an overview is given of market interventions that have aimed to facilitate the local CDM market. One of these measures is the establishment of the provincial CDM centers. These centers have been established by the Chinese government in collaboration with foreign donors in order to steer carbon investment towards previously underdeveloped provincial CDM markets. This can be seen as an attempt to adjust the unequal regional distribution resulting from the diffusion of CDM projects by state intervention. This overview of how the CDM centers are embedded in the Chinese CDM market serves as a background to four case studies on provincial CDM centers.

CDM market development

Although the CDM had been 'invented' as an innovative financing mechanism at the international level by the signatories to the Kyoto Protocol, CDM markets at the host-country level were not emerging spontaneously. Like the demand markets of the Annex I countries of the Kyoto Protocol, CER supply markets in developing countries are political creations that need to be actively brought into being. Creating a new commodity is just the first act in a long play on the international stage of climate governance. Creating functional national carbon markets in CDM host countries is vital for the implementation of the new mechanism. However, many host countries still face the problem of having a high potential market for the CDM, while at the same time these markets are either not (yet) in existence or are in an infant state, so that the ability to engage in CDM business is limited (Nondek and Niederberger, 2005). In this case, interventions by public actors can help to improve market conditions for private actors.

Since the launch of the CDM, China has been seen as the country with the largest potential for a broad range of possible CDM projects (Michaelowa et al., 2000; Hu and Zheng, 2005; Zhang, 2000; World Bank et al., 2004). China is expected to generate at least half of the total CERs up to 2012 (World Bank et al., 2004: xxxvii). China's large potential for the CDM is based on its increasing GHG emissions, its greatly expanding needs for power generation, and its energy efficiency improvement targets, all of which offer various opportunities for large-scale CDM projects. Based on the analysis of marginal abatement costs, the World Bank concluded in 2004 that the power sector could account for about 50 percent of total CDM potential in China, while the steel and cement sector would have the potential to contribute about 10 percent, but they underestimated the huge reduction potential for industrial gases such as HFCs and PFCs (World Bank et al., 2004: xxxix).

China made a slow start on the CDM market in 2005 (Zhang, 2005; see also Figure 2.1). In 2007, China became the world leader on the carbon market, claiming 53 percent of potential CER supply until 2012 to be of Chinese origin (World Bank, 2008: 26). China could now consolidate its top position and claim 40 percent of all CDM projects in the pipeline and a 54 percent market share in terms of transacted CER volume (UNEP Risoe, 2010).

The CDM project portfolio in China shows that a majority of CDM projects are renewable energy projects, while almost one third of CERs are generated by a handful of industrial gas projects (see Figures 2.2 and 2.3).

There is widespread skepticism concerning the quality of CDM projects from China. Most often these projects are criticized for their limited contribution to sustainable development, lack of additionality, and other methodological weaknesses (Haya, 2007; Castro, 2008). There was a period in 2009/10 in which the CDM Executive Board rejected several Chinese wind-power projects on the basis that these would not be additional (He and Morse, 2010). In total, while about 13 percent of Chinese CDM projects were eventually either rejected by

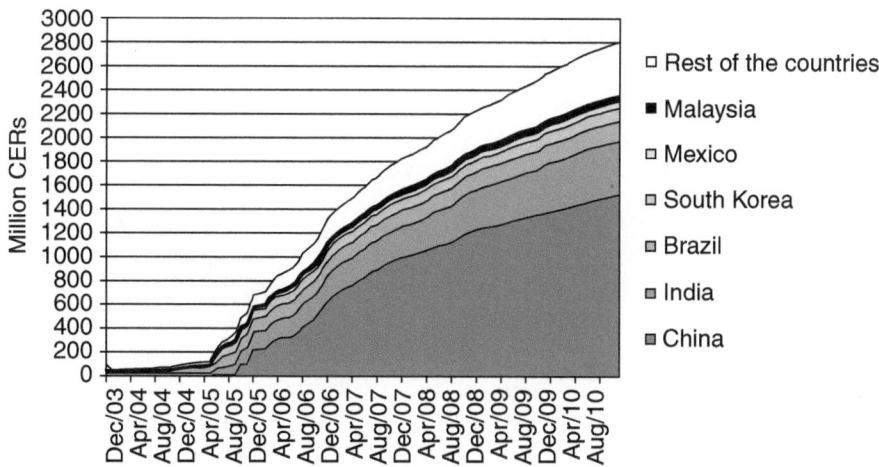

Figure 2.1   China's share of the global CDM market in terms of total expected accumulated 2012 CERs

Source: UNEP Risoe, 2010.

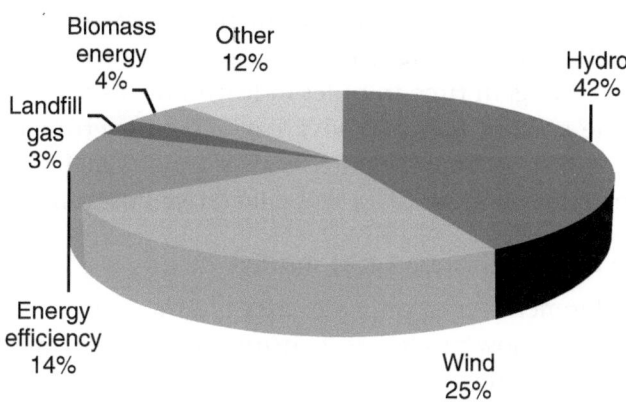

Figure 2.2   Breakdown of CDM project types in China

Source. Based on UNEP Risoe CDM/JI Pipeline Analysis and Database, December 1, 2010.

the EB, terminated, or withdrawn (349 out of a total of 2,633), many projects have had to undergo a review process (57 out of 2,633) (UNEP Risoe, 2010).

Reasons for the slow take-off of the CDM market in China in 2005 included a lack of awareness of how the CDM functions and what kind of business opportunities it offers, but also a skeptical attitude towards the CDM on the part of the Chinese government. The government feared

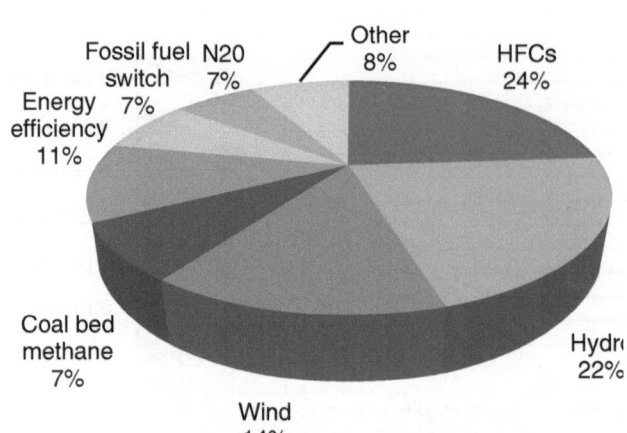

Figure 2.3   Share of CERs by CDM project type in China
Source: Based on UNEP Risoe CDM/JI Pipeline Analysis and Database, December 1, 2010.

that official development assistance (ODA) would be diverted to the CDM, and suspected the CDM of being an invention that would allow developed countries to forfeit their GHG emission reduction obligations. The 'Interim Measures for the Management of CDM Project Activities' (NDRC, 2004) were thus vague in their specifications. Chinese policy-makers wanted to gain time in order to learn how to regulate the CDM market to the greatest national advantage. The government's attitude towards the CDM became supportive after other CDM host countries had demonstrated that the CDM was a lucrative business.

Who's who on the Chinese CDM market

Market development is not only a matter of how new products are dif-fused, but also of how producers, consumers, and their intermediaries and regulators are able to interact.

The highest political institution responsible for climate change issues in China is the 'National Leading Group on Climate Change' (NLGCC), which was established in June 2007 as a successor to the National Coordination Group on Climate Change Strategy (Qi et al., 2008: 381). This group has political backing at a senior level; it has Premier Wen Jiabao as its director, and is also headed by Vice-Premier Zeng Peiyan and State Councilor Tang Jiaxuan. In order to be eligible as a CDM host country, China essentially follows the international requirements for the establishment of institutions for administrating and governing the CDM. Since 2004 it has established a DNA as an office at the politically highly influential National Development and Reform Commission

(NDRC). The NDRC has the mandate to plan, regulate, and control China's macroeconomic strategy. The NDRC has the responsibility to approve CDM projects and to issue letters of approval.

Although the central government tries to speak with one voice to the international community, the internal policy-making and implementation process is characterized by a high degree of competition for influence between political entities, at both local and central levels (Edin, 2003). This kind of ineffective 'stove-pipe management' has also been identified in relation to CDM regulations and institutions. At the central level, both the NDRC and the Ministry for Science and Technology (MOST) take responsibility for the CDM, and compete with each other for influence on the lucrative CDM market (see Figure 2.4) (Interview 2). Officially, the NDRC is responsible for climate policies as they relate to the CDM, while MOST is responsible for science and technology research concerning the CDM. Both ministries are coordinated by the NLGCC. Foreign donors are required to address the NDRC for policy-related programs, MOST for science and technology–related research, and both for the CDM.

The State Council, the chief administrative authority of the PRC, adopted and issued interim measures for the management of CDM projects that became operational in 2005 – measures that were replaced

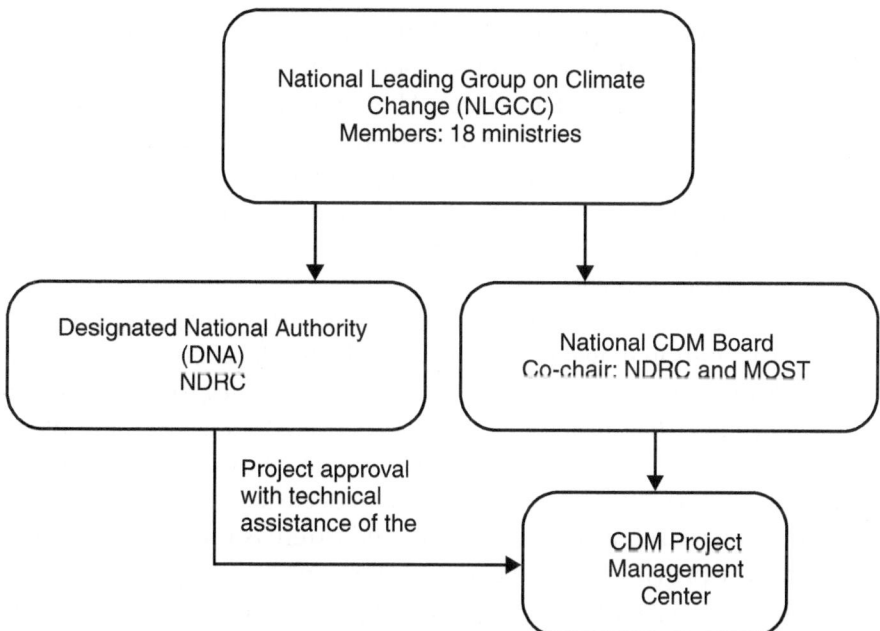

Figure 2.4   Institutions for CDM governance at the national level

by the final CDM regulations in 2007. These specify under what conditions CDM projects can be set up in China and how to win approval for them from the Chinese DNA. The CDM rules set a limit on foreign control of CDM projects by requiring 51 percent ownership by a Chinese company. Although China has not established explicit criteria for sustainable development for CDM projects, sustainable development benefits are expected from projects that fall into the priority areas defined by the Chinese government (NDRC et al., 2005: article 4):

1. energy efficiency improvement
2. development and utilization of new, and renewable energy
3. methane recovery and utilization

These priorities are reflected in the differentiated levies on the CERs' sales from CDM projects: a 65 percent levy is charged on hydrofluorocarbon (HFC) and perfluorocarbon (PFC) projects,[1] N_2O projects have a 30 percent duty, and all other project types incur duty of only 2 percent. The utilization of the international market mechanism of the CDM in the pursuit of China's own political priorities can be described as a 'market capture by the state' (Schroeder, 2009). Revenues from these levies are allegedly flowing into the newly established China Climate Fund (the initial idea of which seems to have come from World Bank advisors), which will provide funding for more sustainable projects such as renewable-energy and energy-efficiency projects and capacity-development activities (ADB, 2006). The national taxation of revenues from an international market mechanism can be seen as a textbook example of how perceived international policy failures can be corrected by state intervention if – and here doubts remain – collected revenues are indeed used effectively for the financing of renewable-energy and energy-efficiency projects.

The bureaucratic structure of the CDM at the local level mirrors that at the national level: the provincial Development and Reform Commissions (DRCs) take care of CDM-related policies, and the provincial Science and Technology Departments (STDs) focus on technology transfer and research (Interview 3). In practice, however, competition continues at all levels, and in many provinces either the DRC or the STD has established itself as the CDM hub, and often as the mother organization for the provincial CDM center (Interview 2) (see Figure 2.5). Only rarely do both departments cooperate in CDM matters. Also mirroring the national situation, provincial governments have started to set up provincial-level 'leading groups on energy saving, pollution reduction, and climate change,' which have the mandate to organize the implementation of national policies and strategies in these issue areas (Qi et al., 2008: 382).

Of the 1,286 project developers active globally, 216 are active in China and one operates in Hong Kong (see Figure 2.6) (UNEP Risoe, 2009). Of the active projects developers, 160 are CDM consultancies, while the remaining companies focus on other business areas. These can be power-generating companies, which act as simultaneously as project owners and project developers. No specific qualification requirements for CDM project developers are issued either by the EB or by the Chinese DNA. It is therefore sometimes difficult for inexperienced project owners to decide which consultancy to cooperate with in CDM project development.

Of the 33 Designated Operational Entities (DOEs) operating worldwide, 26 are active in China. Of the CDM projects registered or in the process of being registered in China in May 2010, 29 percent had been validated by Det Norske Veritas (DNV), another 24 percent by TÜV Süd, 10 percent by TÜV Nord, and the remaining 37 percent by 23 other international DOEs (see Figure 2.7). The timely delivery of DOE services is currently subject to a bottleneck in China, because the DOEs operating in China lack sufficient qualified staff and suffer from a work overload. One way of relieving this might be the accreditation of several Chinese DOEs.

The CDM-related financial community in China is dominated by private foreign banks that act as CER buyers and traders, such as

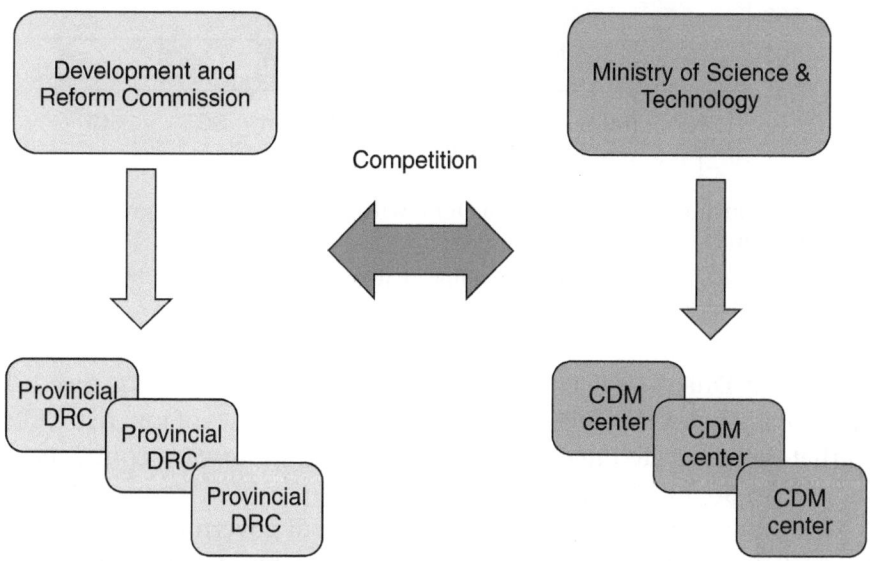

Figure 2.5   Competition between different state institutions

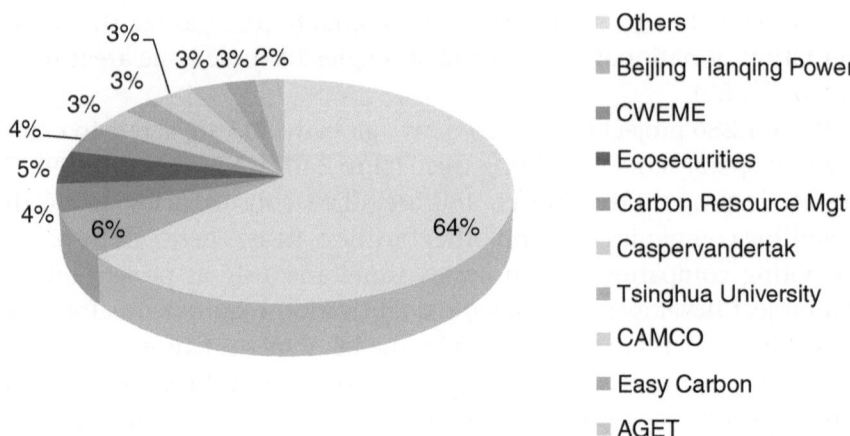

Figure 2.6 Share of total projects developed by different CDM project developers operating in China

Source: Based on UNEP Risoe CDM/JI Pipeline Analysis and Database, December 1, 2010.

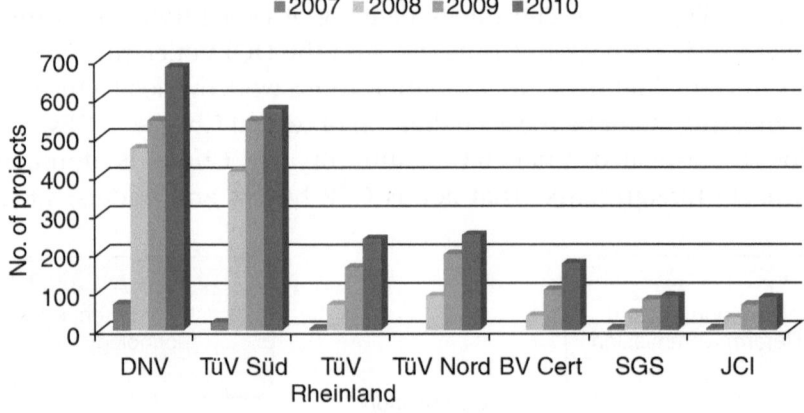

Figure 2.7 Business development of DOEs with a large number of project validations in China

Source: Based on UNEP Risoe CDM/JI Pipeline Analysis and Database May 2007, May 2008, May 2009, and May 2010.

the British-Dutch ING Bank, and private funds, such as the Swedish Tricorona. Their main objective is to buy large tranches of low-risk CERs, so that most private buyers are only interested in forward contracts on more than 50,000 CERs and do not provide upfront investment loans.

Most interesting is the ambiguous position of governments of Annex I signatories to the Kyoto Protocol. On the one hand, national governments are the parties to the Kyoto Protocol, and thus retain the right to participate and vote in negotiations and to set and modify the rules of the Protocol. On the other hand, the governments of Annex I countries have agreed on binding GHG emission targets, and thus have a strong interest in buying CERs from CDM projects as one means to fulfill their commitment. As both rule-setters and a player in the Kyoto game, many governments thus have a variety of interests when deciding about CDM capacity development programs in CDM host countries.

CDM in the provinces

The following section introduces the different framework conditions of China's provinces. This information provides the background for an analysis of why some provinces do better than others in CDM project development. It also explains the motivation for state intervention to steer greater investment into the underdeveloped western provinces.

China's provinces: regional gaps in development

In discussing local CDM market developments in China, it should be kept in mind that China resembles a subcontinent with provinces that match European countries in size, population, and productivity. Administratively, China is divided into 22 provinces, five autonomous regions, four municipalities, and the two special administrative zones – Hong Kong and Macao.[2] An established way of differentiating between provincial-level units is to distinguish between provinces of the eastern coast and those of the western interior.

The eastern provinces are the powerhouse of China's industrial growth. They have a high concentration of export industries, receive the bulk of foreign investment, and have a fairly well-educated, urban population. Industrial energy demand and rising living standards resulted in the coastal region accounting for 70 percent of the energy demand growth in China between 1996 and 2005 (IEA, 2007: 403).

In contrast to the coastal boom provinces, the inland provinces account for only 16.9 percent of national GDP, and their per capita income reaches only 60 percent of the national average, although they take up 71.5 percent of the total area of China and are home to 28.7 percent of China's population. The state revenues in these provinces are only 17 percent of the national average, while their expenditures represent 24.9 percent (National Bureau of Statistics of China, 2005: 41). Although China is the world's second-largest FDI recipient (IEA, 2007: 243), with a total inflow of around US$108 billion (IEA, 2007: 246), only 12.8 percent of all FDI was invested in China's western and central regions, whereas 84.56 percent was invested in the eastern provinces. If provincial development is based on the individual provinces' annual GDP growth rates, China's regional gap in economic development is doomed to grow larger.

Utilizing the CDM for western development

One of the greatest political challenges for China is therefore to level out the development disparities between prospering coastal regions and the interior. In order to narrow this gap, Jiang Zemin, President of China from 1993 to 2003, launched the Western Development Strategy (*Xibu da kaifa*) in January 2000 as a cornerstone of the Tenth Five-Year Plan (2001–05). The strategy envisages the development of the inland regions up to the level of a 'well-off society' (*Xiaokang shehui*) until the mid-21st century (Lai, 2002; Lahtinen, 2005: 24).

The CDM is also envisaged to contribute to the government's Western Development Strategy because it promises to draw foreign investment into the extension of the energy infrastructure, while simultaneously contributing to sustainable development. Environmental protection has to play a major role in the economic development of the western provinces because China's western regions have abundant natural and mineral resources. Their usage will increase the pressure on the local eco-system, among other impacts, increasing GHG emissions. The Chinese government thus envisages a transfer of clean technologies to the western provinces, and regards the CDM as a potential channel for technology transfer and investment from China's eastern provinces and from foreign countries. Revenues from the CDM can in turn help to overcome the budgetary constraints of local governments, and can provide additional financing opportunities for companies interested in developing projects with GHG emission reduction potential (Interview 14).

The CDM potential of China's provinces

The eastern provinces: source of the bulk of
China's industrial GHG emissions

Not surprisingly, the coastal region also accounts for most of China's GHG emissions. Their annual average growth rate of 13 percent between 2000 and 2005 was higher than China's average annual rate of 11 percent. The coastal GHG emission growth rate is projected to decline only slowly, to 5.2 percent by 2015 and 3.3 percent thereafter (IEA, 2007: 418). Home to China's chemistry and manufacturing industries, to its large cities, and to its agricultural heartland, the eastern provinces have good CDM potential for HFC, energy efficiency in industry, landfill gas, and N_2O projects – all of which are CDM project types that yield comparatively high levels of CERs.

The western provinces: blessed with good energy resources
The western region holds huge reserves of energy and mineral resources which could help to turn these regions into China's industrial and energy powerhouse (Lahtinen, 2005: 25). The region contains 64.1 percent of China's coal reserves, and the its coal bed methane potential accounts for 57.8 percent of the country's. Renewable energy resources are also abundant in the western provinces: Inner Mongolia has the country's best onshore wind sites, the Tibetan plateau has China's longest periods of daylight, and the provinces connecting the Himalaya mountain range with the flat eastern regions have significant hydropower potential. This makes the western region of interest for various CDM project types: energy-efficient coal-fired power production, coal bed/mine methane, hydropower, wind power, biomass energy and biogas projects. The renewable energy projects tend to have better-than-average sustainable development benefits, though they yield fewer CERs.

The regional gap in CDM project development in the early CDM market

Despite their good energy resource allocation, the western provinces made a slow start in the CDM market, while the eastern provinces hosted the first CDM projects in China. In general, the first movers on the CDM market were the coastal provinces, which had already received a significant amount of FDI. One notable exception was the landlocked Autonomous Region of Inner Mongolia, which was home to the second CDM project in China and has done well in CDM project development. One representative of the NDRC explains Inner Mongolia's ability to realize its CDM potential early on through its good project development capacities, as the region is home to several local wind-power companies that already had considerable experience in how to develop wind-power parks in collaboration with foreign companies (Interview 1).

Comparison of the western provinces' CDM potential with the number and CER volumes of CDM projects developed in 2005 reveals a clear paradox. Despite their good potential, the adoption rate of the CDM in the western provinces was low at the beginning of the Chinese CDM market. This paradox between high potentials and low adoption rates has also been observed in other markets (for example, for energy-efficient technologies – Jaffe et al., 2002: 48; Shama, 1983). With the progress of the Chinese CDM market, the situation has changed fundamentally: the western provinces caught up in 2007, at least in terms of the number of CDM projects – although, in terms of CERs generated, large HFC and N_2O projects on the eastern coast still form the basis of a geographical gap. Also, the correlation between CDM projects and GDP per capita has reversed for the years 2008 and 2009: now the underdeveloped western provinces, and not the prosperous coastal provinces, hosted the majority of CDM projects (see Figure 2.8).

CDM barriers and needs for provincial CDM capacity development

For a better understanding of why state intervention can be necessary for facilitating markets, I outline here the barriers that hampered CDM market development in the early phases but seemed to have been overcome in the later phases. CDM project development is a challenging task because several barriers have to be overcome: potential project owners may not have heard about the CDM before, and even if they have, they may lack a clear understanding of it, or even distrust it – and they may not know whether their project is CDM eligible. Similarly, provincial government officials are often unaware of the CDM and unsupportive of CDM project development in their province. Even if they understand and appreciate the CDM, it might nevertheless be

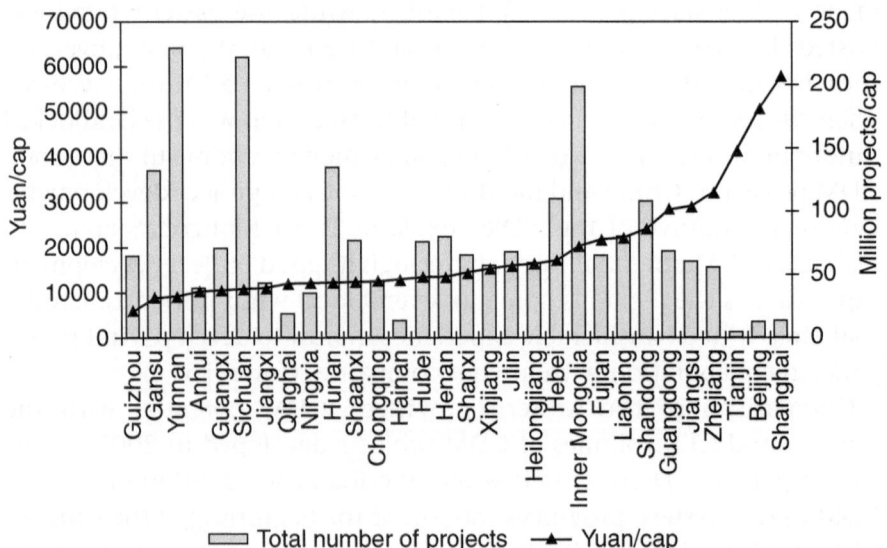

Figure 2.8   Total number of projects and GDP per capita in Chinese provinces
Source: UNEP Risoe, 2010.

hampered by an inefficient bureaucracy or a general lack of coordination among government institutions that are responsible for it. In cases where awareness among potential market participants is sufficient, CDM project development may still be hampered by an inability of local project developers to develop CDM projects that achieve DOE validation, host-country approval, and EB registration. Lack of experience in cooperating with foreigners can also make signing Emission Reductions Purchase Agreements (ERPAs) with foreign buyers difficult, as no contacts with foreign companies are available.

National-level CDM-related barriers are discussed in the literature (Ellis and Kamel, 2007), but they can also be observed at the provincial and local levels. Barriers can arise in different phases of the CDM project cycle; they might vary over time and between locations; and each combination of barriers can have special effects. They also imply different cost and time requirements to be met (Ellis and Kamel, 2007: 6). For the provincial CDM markets in China, the following four main categories of such requirement can be identified: (1) awareness, (2) expert knowledge and data availability, (3) governmental support, and (4) access to finance.

1) Awareness and trust about climate change and the CDM

Awareness of the CDM is needed at various levels within a host country. At a minimum, the following stakeholder groups should be aware of the CDM (Ellis and Kamel, 2007: 29):

- policy-makers at different government levels, who are responsible for issuing CDM-related policies: ministers, heads of departments, provincial governments, and city municipalities;
- persons at the project level, whose task is to identify and develop CDM projects: rural electrification practitioners, consultants, and academics;
- financiers, who can provide loans to CDM projects: local and national bankers, loan officers, and individuals working in local financial intermediary organizations.

At the very beginning of the CDM market in China, the level of CDM awareness and knowledge was low among potential CDM market actors outside Beijing. Chinese market participants neither believed nor understood the CDM. It was commonly referred to as a 'cake fallen from the sky,' because people did not believe that they would be paid by foreign companies and governments for reducing emissions. Prior to the CDM, reducing emissions was seen only as part of a tedious environmental protection effort that limited economic development. Companies preferred to pay the fines of the Environmental Protection Bureaus rather than to purchase emission-control equipment. The CDM contributed to changing this attitude (see Heggelund and Ying, 2008: 12). But although the CDM hit the headlines not only in the international media, but also in China, many potential project owners were still badly informed as to what exactly the CDM meant, and what kind of projects would be eligible. As a study by the China Council on International Cooperation for Environment and Development (CCICED) shows, awareness of climate change and the local potential for the CDM was almost absent in the western provinces:

> As a result of problems accessing information, enterprises and the people of the Western provinces know little about climate change and CDM, and have no deep experience of the opportunities offered by implementing CDM projects. Though the consensus is always emphasizing sustainable development, most people, especially the decision-makers of economic organizations, still put their main concentration on economic development. Some think that environmental protection contradicts economic development. Improved access to information is required to show that economic progress and environmental protection can – when properly managed – be achieved simultaneously. (CCICED, 2001: 16)

Although this situation has improved considerably for industry sectors in which the CDM is now frequently used, there are still many sectors in which CDM knowledge and capacity remain very low, and initiatives for project development are not being taken up (Interview 27). Judging from this situation, one can conclude that there is a need to raise awareness of the CDM and of project eligibility. This need for the building of awareness of the CDM persists today in regions and sectors that are not at the forefront of CDM project development.

2) Expert knowledge and data availability

In the early Chinese CDM market, expertise on CDM project development was lacking at the local level. Most CDM projects were developed by international consultancies or by Chinese experts from Beijing universities and research institutions. These dozen experts were not enough to tap the full potential of CDM projects in China. Local knowledge and personal networks are important for identifying projects and taking local circumstances into account. This lack of expert knowledge at the provincial level can severely hamper the ability to identify, develop, and register CDM projects, and to find buyers for the CERs (OECD, 2007a: 6 identifies this barrier for market development in general; Del Río, 2007: 1,367 discusses it specifically in relation to the CDM). The coastal provinces were slightly better off in this respect, because their companies have abundant experience in how to attract foreign investment and deal with foreign companies, because they usually have established networks and contacts with foreigners, and because their foreign language skills are better.

The lack of data on GHG emissions was also a barrier for the early Chinese CDM market. For the establishment of baselines and for the calculation of expected CERs, historical and current data on industrial activities and emission factors need to be available for the different CDM sectors and for China's various regions. The more accurate the available data are, the more easily projects' risks can be calculated and assessed for the project developer (Ellis and Kamel, 2007: 15). Small companies, in particular, cannot afford time and effort to gather relevant CDM information by themselves. In this kind of situation, public support in collecting and analyzing data and making them publicly available would be needed (Ellis and Kamel, 2007: 15). Unfortunately, sufficient and reliable data were not available for CDM project planners in the early market, which deterred project developers and buyers. This lack of adequate data was due to the lack of technical capacities to monitor, collect, and manage data about GHG emissions, and to the lack of coordination and willingness on the part of various governmental entities to consolidate and share these data transparently with the public.

This results in the need to strengthen local capacity and knowledge about CDM procedures and project planning, and to interact with foreign business partners. It would also be necessary to improve capacities to collect, manage, and share data on GHG emissions and prepare pre-feasibility studies. A one-stop shop for consolidated information about the provincial CDM situation – its potential, special regulations, responsible government entities, business counterparts, and contacts to foreign companies – would be helpful in overcoming this information deficit and convincing local companies to get involved in the CDM.

3) Governmental CDM support

Although the CDM project approval process rests with the central government, local governments are important actors in terms of approving the project as such (besides the financial CDM component) and in cooperating in the DOE validation process. It is of course always beneficial for market participants to keep a good communication with the local government and to keep it informed about CDM projects and ensure its cooperation in project implementation. After all, it is the local DRC which has to give its permission to all non-CDM related matters of infrastructure projects and which needs to cooperate for DOE validation. Thus, also at the provincial and local level, the support of the government is needed for CDM market development. Crucial for investors is the existence of foreseeable and enforceable CDM-related policies that reflect the supportive attitude of the government.

It is commonly understood that predictable and simple policies and their effective implementation are a precondition for attracting foreign investment (OECD, 2006b). Inconsistent policy implementation exacerbates uncertainty, and therefore increases time and resources needed to interpret ambiguous legislation, and thus increases the risk of contractual non-compliance (Ellis and Kamel, 2007: 19). As the various types of eligible CDM projects are implemented in a variety of sectors and regions, public support for the CDM in each of these areas is relevant. The need to strengthen local governmental support and local state capacity for the CDM thus becomes clear.

4) Access to project finance

Especially in the early CDM market in China, access to finance was a major problem for CDM projects. For-profit CER buyers usually refrain from providing upfront finance for CDM projects. Local project owners thus have to turn to local financiers for upfront investment. For the local financiers, a thorough understanding of CDM requirements would be needed in order to be able to assess a project's eligibility for loans based on its cost-benefit calculations. Unfortunately, it has not been possible for this knowledge of CDM opportunities and risks on the part of local financial institutions to be taken for granted. The lack of access to upfront project finance can thus be a severe barrier for CDM project realization (Ellis and Kamel, 2007: 29).

Chinese banks are generally hesitant about lending money to private companies, especially for small projects. This phenomenon has been identified as one of the main problems for China's private sector growth (Tsai, 2001; Firth et al., 2009). Banks' reluctance and lack of interest in lending to private companies could also be observed on the CDM market in 2005. Local banks hardly knew about the CDM, and were thus unwilling to take it into their lending considerations. Despite official backing of the CDM by the Chinese government, the situation had not improved much by 2011. Access to project finance remains a major bottleneck to CDM project realization.

All of these barriers pose challenges to actors who are interested in becoming active in the CDM market. As table 2.1 shows, each market actor group faces specific barriers and capacity development needs. If these barriers cannot be overcome, potential project participants are likely to lose interest in the CDM because they are unable to judge the risks it involves. In many CDM host countries, successful demonstration projects, which might have demonstrated that engaging in the CDM is a lucrative business, were missing at the beginning of the carbon market. In such situations, public interventions can facilitate the adoption of the CDM by helping potential project participants to overcome the barriers. The right-hand column of Table 2.1 suggests several capacity development measures that might be helpful in overcoming the CDM deficits within each actor group. As the following section on capacity development will show, not all actor groups and their needs are equally addressed and targeted by CDM capacity development programs in China.

Table 2.1   Overview of capacity needs in the early Chinese CDM market

Actor		Areas of inadequate CDM capacity	Possible capacity development activities
Citizens		Awareness and knowledge of climate change and CDM	Educational campaigns
		Vulnerability to climate change	Adaptation measures
		Knowledge of low-carbon lifestyles	Educational campaigns
Public actors	National government	Skeptical attitude towards CDM and fear of losing 'national resources'	Successful demonstration of projects in other host countries
		Trust of private (foreign) consultancies	Quality standards/ accreditation
	Provincial government	Awareness and knowledge of climate change and CDM; Enforcement of national climate-related policies	CDM training, workshops, and conferences
		Political priorities/policies for mitigation and adaptation	Provincial climate change program
		Enforcement of national climate-related policies	Strengthening of state capacity or outsourcing to private sector

Continued

Table 2.1　Continued

Actor	Areas of inadequate CDM capacity	Possible capacity development activities
Private actors　Project owners	Awareness of CDM opportunities	Conferences, demonstration projects, media coverage, feasibility studies, project screening
	Availability of one-stop shop for CDM background information and advice on CDM procedure	CDM centers, CDM publications, workshops/training
	Availability of relevant data for the project design document (PDD)	Provision of data by DNA
	Capacity to monitor own GHG emissions and CDM activities	'On the job' training
	Ability to sell CERs: • Information on potential buyers • Network with potential buyers • English language skills	Match-making activities, setup of networks/ internet platforms
Project developers/ Consultancies	Capacity to develop PDD according to requirements: • Use of CDM methodologies • Proof of additionality and sustainable development benefits • Adequate stakeholder consultations • Understanding of (inter) national CDM approval process • Understanding of markets and CER prices • Familiarity with ERPAs or similar internationally accepted contracts • Setting up monitoring plans	Learning-by-doing, training
Financial institutions	Awareness of CDM opportunities and their financial potential and risks	Training
	Knowledge of secondary carbon market and CER trading opportunities	Training
Others: NGOs, researchers, media	Knowledge about CDM procedures	Publications

Capacity development for the CDM in China

A great deal of information has been gathered about institutional capacity building for climate protection at the national level, but measures to extend countries' capacities for climate governance to the sub-national level have begun only recently. While capacity development needs at the local level might initially be the same as on the national level, experience has shown that donors tend to focus on the national level first. This can be explained partly by the fact that local capacity development poses strategic challenges to donor organizations, as these tend to operate at a national level but lack access to information and local networks (OECD, 2000: 12). Another reason why CDM capacity development programs usually take place in cooperation with entities of the Chinese government is, on the one hand, the partnership approach of most donor agencies, and, on the other, the Chinese government's suspicion and hesitancy about foreign interference in its internal affairs.

Capacity development measures for the CDM market in China can be divided into three phases. Before the Kyoto Protocol came into force in 2005, projects naturally focused on the assessment of the Chinese potential for the CDM by analyzing most promising sectors for early and relatively straightforward CDM projects and by drawing policy recommendations for the Chinese government on how to set up an effective institutional framework (see World Bank et al., 2004). In a second phase, capacity development projects focused on developing demonstration projects for the CDM. In the third phase of engagement, capacity development projects were to support CDM market development by accessing new sectors and new regions and by assisting in the development of new CDM methodologies.

The main document outlining the national priorities of the Chinese government in relation to climate change is its National Climate Change Program. Within this program the Chinese government also aims to strengthen local capacities for mitigating and adapting to climate change. This capacity development is intended to include the establishment of a regional administration system and local expert groups for coordinating climate-related activities. It also envisages strengthened coordination between national and local governments for climate change policies and measures (NDRC, 2007: 56). The local Development and Reform Commissions are the responsible units for their implementation. Their main task is to advise the provincial government on climate-related policy and on how to implement objectives of the National Climate Change Plan locally.

The Chinese government also launched a Provincial Climate Change Program under the supervision of the NDRC, with the financial support of US$750,000 from the Italian government and technical support from the World Bank China Office. The program's objectives are, first, to demonstrate the positive contributions of China in climate protection to the international community; second, to identify policy challenges in relation to climate change and sustainable development; and third, to improve the local capacity to deal with climate-change issues. Pilot programs have been completed for Hubei, Jilin, Shaanxi, and Yunnan provinces. In May 2008, all Chinese provinces were finally asked to develop their provincial programs (Hunan CDM Center, 2008). This directive and the central government's 'energy saving and emission reduction' campaign, plus the opportunity to raise additional revenues from CDM projects, motivated local government officials to take a positive stance towards the CDM (Qi et al., 2008).

The idea of setting up provincial CDM centers

This section clarifies how the idea for setting up provincial CDM centers was born and diffused among 27 provinces. Until now, the Administrative Center for China's Agenda 21 (ACCA 21), under the supervision of MOST and with financial support from various foreign donors and organizations, has supported 27 provincial governments in setting up 'provincial CDM centers.'

The interests of involved actors

All groups of actors interested in developing provincial CDM markets share one common interest – the full utilization of the provinces' CDM potential – although for different purposes. In essence, the central government wants a check on private, especially foreign business actors, and wants to deploy the CDM to improve the sustainable development of the western provinces; local governments strive to fulfill national targets, and are keen on extra revenues from CDM projects; multilateral organizations hope to see carbon finance used for achieving MDGs; Annex I governments and companies strive to purchase CERs; and NGOs see the CDM as providing an additional opportunity for raising environmental awareness.

The Chinese central government pursues several interests by supporting provincial CDM centers. Generally, the government supports the provinces in realizing their full CDM potential, and was encouraged by foreign experts to do so by ensuring the quality of CDM projects developed in remote provinces. As early as 2002, the CCICED warned that many Chinese companies, especially in the western provinces, would not be trustworthy enough to act as CDM cooperation partners for foreign buyers, and advised the central government to initiate the CDM project development by governmental means, and to support local market development:

> Not all the Western provinces are suitable hosts for CDM projects. To implement the CDM not only requires local firms to have financial and technological capabilities, but also requires that they are able to gain project information, select technologies, negotiate on projects, manage operation, and calculate emission reduction credits. This is one side. On the other side, if foreign governments and companies do not have sufficient confidence in local firms, CDM projects will be difficult to implement even if they have some capabilities. Therefore, in the initial stage of CDM cooperation in particular, the central government of China should lead and supervise the implementation, to select some appropriate Western provinces that have relatively sound financial resources [and] powerful firms with international horizons and the ability to cooperate. (CCICED, 2001: 17)

State intervention was not only a means to support local companies and enable high-quality CDM projects, but also a way to monitor the business practices of foreign and Beijing-based companies. One central reason for setting up CDM centers that are closely affiliated with the government is that they help the central government maintain control over market forces and the activities of foreign companies, as one government official explained:

> The overall goal of CDM centers is to support provinces in realizing their full CDM potential. In the beginning of the CDM market, consultancies mushroomed in Beijing only. We thought that within two to three [years], the CDM should not depend on consultancies only, so the idea for the Provincial CDM centers was born. ACCA 21 of MOST presented the idea to several donors and asked for international funding. (Interview 1)

The government's wish to maintain control especially over foreign CDM market participants was confirmed in an interview with a foreign CDM project developer:

> One reason for the establishment of provincial CDM Centers is also the goal of the Chinese government to keep control of the CDM

market. They were suspicious and disliked some 'bad' consultancies that either used fraud documents in PDDs or charged [a disproportionally] high service fee. (Interview 18)

Another reason given for the establishment of the provincial CDM centers was the expectation that realizing the CDM potential of the western provinces would contribute to the Western Development Strategy:

Building provincial CDM centers is part of a strategy to utilize the CDM for China's Western development. Of course, compared to the total investment needs, the contribution of the CDM is small. However, besides project financing by CERs, local CDM centers also get other local benefits. For example, they have more opportunity to cooperate with the outside world, awareness of climate change issues increases, and human resources are developed, which is an important task. (Interview 1)

This motive was also acknowledged by a representative of an international consultancy operating locally, who stated that the central government had been quite concerned about climate change and had taken up related measures, though these would need time and effort to spread to the provinces, where not much had happened yet. Another motive for setting up provincial CDM centers would be the interest in raising additional revenues from the CDM (Interview 52).

Concerning the tasks of the provincial CDM centers, a representative from MOST explains:

The writing of PDDs is only one aspect of the provincial centers' work. Other objectives are dealing with international organizations such as donors and to establish contacts with international companies that buy CERs. They are supposed to provide consulting services to project owners, organize training for local companies, and increase understanding of the CDM. Many potential project owners at first did not believe the CDM. They did not believe that someone was going to pay for reduced GHG emissions. The centers are also supposed to provide background information on climate change, and thus increase climate change awareness. They should work independently and get the ERPA contract from project owners. The centers receive revenues from their consulting services and are able to develop networks. (Interview 1)

An interview with a Chinese project owner revealed different motivations. According to him, the establishment of provincial CDM centers was a move by MOST to strengthen its presence in the provincial CDM markets, thus gaining ground in relation to the NDRC:

> The first CDM center was set up in 2003. At that time, MOST, not the NDRC, was responsible for the CDM. [MOST officials] wanted to have local departments to support them in their work. When in 2004/05 the responsibilities [for the CDM transferred] to the NDRC, China already had in total over 20 [MOST-affiliated] departments having a say on the CDM. Now the NDRC is responsible, and the Energy Departments of the Provincial DRCs take the local lead. So, the CDM centers under the MOST are a product of history. At this time, NDRC has no provincial-level CDM counterpart. But not all the existing CDM centers are a part of the provincial Science and Technology Department; sometimes they are a 'Shiye Danwei' under the MOST, but they are also often an individual's private company. The position of the CDM centers is very different – quite confusing – for each province. Sometimes it is a part of MOST; sometimes it is not. (Interview 47)

Five main reasons can be identified as to why the central government was interested in facilitating provincial CDM markets by setting up CDM centers: first, to strengthen local companies and ensure high-quality CDM projects in remote provinces; second, to keep a check on foreign CDM consultancies and prevent them from capturing rents; third, to deploy the CDM for pushing sustainable development in the western provinces; fourth, to win a share of the CDM market for public entities; and fifth, to win influence for the MOST at the local level.

Donor interests

One major factor determining a donor's interest is its domestic demand for CERs. This motivation was expressed, for example, by a representative of the Canadian government, which supported China's first provincial CDM center in Ningxia:

> The Ningxia project was a very small one, a pilot project, and then it developed into a much larger project, because at that time we realized that Canada was going to have a purchasing agency looking to buy CDM credits. And this is one way of focusing our efforts in one particular area that potentially could lead to the development

of PDDs and CERs that the government could consider purchasing. (Interview 5)

But Canada is an awkward case: it was one of the first countries to support CDM capacity development projects at the provincial level, and even expanded its initial project to four other provinces. Although Canada has been well exposed in terms of capacity development for the CDM provincially, it has not yet bought any CERs directly from a CDM project – not even from its 'own' provinces – but has so far only bought CERs through international funds. Canadian internal political changes (from Kyoto supporter to Kyoto 'laissez-faire,' after a change of the Canadian government in 2006) may explain this situation. The same interest in buying CERs from the province in which they support a CDM center can be observed for other donor countries. The UK and France have sent trade delegations to the provinces of their CDM center programs: the UK Emission Trade Team visited Guizhou province, and a delegation of ten French enterprises, CER buyers, financiers, and companies selling equipment for CDM projects has visited the four south-western provinces where Agence Française de Développement (AFD) has a CDM center program. Although UK companies have a 38 percent share of CERs sold from the Guizhou province, this is not a result of the program's activities, as it merely mirrors the percentage of the UK's CER purchase from the Chinese total (UNEP Risoe, 2009).

This need for CERs from the buyers' side is matched by an interesting offer from the Chinese government. Governments that support provincial CDM centers can gain the 'first right of purchase' for CERs for a limited amount of time. According to several interviewees, donor agencies and/or their domestic companies (depending on who has the demand) are unofficially entitled to enter into ERPA agreements with CDM projects that were sourced and developed by the CDM center of 'their' province:

Many donors have the first right to choose from PIN/PDDs developed in 'their' province within a certain time period. If they are not able to find a buyer from their country within that period, the CERs can be freely sold on the market. UNDP does not have such a first right of purchase negotiated with MOST, since their Global CDM Carbon Fund (just launched in June) [so far has] little demand, so that they do not see the need. (Interview 1)

The origin of the idea

The very first idea of establishing a provincial CDM center came from the local level. Zhang Jisheng, director of the Ningxia CDM center, claims he was the first to come up with the idea as early as 2003. Apparently out of his realization of the high CDM potential of Ningxia, he talked to MOST, which eventually got in touch with the Canadian DFAIT to solicit support in setting up China's first CDM center in Ningxia in October 2003 (Interview 5). The Chinese-Canadian capacity development program focused on the western provinces of Ningxia and Gansu, as these lacked CDM capacity and had expressed their need for support. Eastern provinces were thought to be able to tap their CDM potential without external support, due to their good human resource situation and their prior experiences with Western business partners (Interview 1). After the successful establishment of CDM centers in Ningxia and Gansu, MOST diffused the idea to other provinces, and asked an additional 25 provincial governments to choose provincial-level institutions that could serve as CDM centers.

Surprisingly, it was not clear from the beginning even that MOST should host the provincial CDM centers, because the United Nations Conference on Trade and Development (UNCTAD) started to finance a 1.5-day CDM training program in September 2005 in Xiamen City for the provincial investment promotion agencies to become the local CDM hubs (Policy Solutions and CSEND, 2005). Apparently, this program was phased out without a successor. The main reason for that was probably that MOST and NDRC – and not the Ministry of Commerce (MOFCOM), which oversees investment promotion agencies – were able to take the lead on the CDM in China.

Mandates of the provincial CDM centers

The provincial CDM centers have a diverse range of mandates that include CDM promotion, capacity development, policy advice, research, and CDM project development. There is no central directive relating to CDM centers' objectives, and thus provincial CDM centers have some freedom in choosing their own focus. CDM centers usually concentrate on a few core activities. Representative for many is the mission statement of the Sichuan CDM center, which describes its tasks on its website as follows:

> The center's main tasks include the buildup and gradual improvement of the whole province's CDM project policy framework;

the natural resource management and establishment of a public-private model of CDM project development; the establishment of a list of technicians and CDM specialists; the training of CDM intermediary service agencies; the continuous expansion of the CDM market; and the establishing of international exchange and cooperation networks with many actors and various CDM project developers.

According to a news clipping of the main Chinese news agency Xinhua, CDM centers' mission is 'to raise the awareness of the local officials, enterprise managers and the public on CDM, train CDM talent, promote CDM sample projects, and introduce advance technology and investment from abroad so as to help the local areas increase their capacity in responding to global climate change' (China CSR.com, 2007). Table 2.2 gives an overview of what objectives the individual CDM centers state on their websites, if they have one.

The core objective of all CDM centers is to offer project development services. In addition, the mandate of these centers includes disseminating information to local stakeholders and improving CDM capacity at the local level. Individual centers can have additional activities in their portfolio.

Table 2.2 Mandates of selected CDM centers according to their websites

	Research	Policy advice	CDM market info	Trainings	Project identification	Writing of PINs	Writing of PDDs	ERPA negotiation	Methodology development	CER trading	Other
Guizhou					X	X	X				
Hebei			X	X	X	X	X	X			
Hunan		X	X	X	X	X	X	X	X		
Jilin			X	X	X	X	X	X			
Ningxia	X			X		X	X				
Shandong				X	X	X	X				
Shanxi						X	X	X	X		
Sichuan		X	X	X	X	X	X	X	X		X
Yunnan				X	X	X	X	X			

Source: Websites of the CDM centers

Design and functions of the CDM centers

The provincial CDM centers are always in some way affiliated with the local government; but there is no top-down regulation governing what that relationship should look like. CDM centers can be part of local government, a private company, or a mixture of both; nor are there regulations governing the specific tasks of the CDM centers, or on how much of their total workload their CDM work should take up. Indeed, for many centers, the CDM is only a part-time activity (Interview 1). There is also some competition over which institution may call itself a 'Provincial CDM Center,' leading to a situation in which the Sichuan province has two 'official' CDM centers (Tuozhan CDM Service Center and Hareal Consulting Co., Ltd) which both receive support from AFD (Interview 14). This situation emerged because the selection of the institution to become the provincial CDM center is not carried out by MOST, but by provincial government institutions, which may themselves have conflicting interests. As a consequence, the provincial CDM centers need not necessarily to be attached to the Department of Science and Technology (a provincial subunit of MOST), but can also be part of the Development and Reform Commission (a provincial subunit of the NDRC) or the Environmental Protection Bureau (a provincial subunit of the MEP) (Interview 1).

Currently, 27 provincial CDM centers are established or in the planning stage. Only four local entities – the cities of Beijing and Shanghai, Guangdong province, and the autonomous region of Tibet – have no CDM centers. Setting up CDM centers is seen as unnecessary for the municipalities of Beijing and Shanghai and for Guangdong Province, because these are already home to many private consultancies, and are the most economically advanced places in China. The opposite is true of Tibet, which is an economically underdeveloped province and which as yet has no CDM project, although it would be possible in theory. The 27 CDM centers are supported by a diverse set of six donors, including government agencies of Canada, France, and Japan; multilateral organizations such as UNDP and ADB; and even a private financial institution, the Dutch ING Bank. Arcelor Mittal, a transnational steel company, was supposed to finance UNDP's CDM center program, but withdrew its support after the onset of the financial crisis (Interview 3). MOST, however, has been the implementing agency for all 27 CDM centers, and thus their capacity development programs followed a similar model. There has been no coordination among the different donors about their CDM center programs. Their sole communication point has been MOST.

Despite the involvement of the different donor agencies, and despite their different local institutional configurations, MOST holds the overall program supervision and ensures that all CDM center programs follow a similar set of activities. This package includes the following components:

- *Awareness raising.* The methods of awareness creation used by the CDM centers are very similar, but differ in their timing and scope. All CDM centers are involved in organizing provincial CDM conferences for spreading CDM awareness among local political leaders, industry, and media representatives, and all have published basic information on the CDM.
- *Capacity development.* All CDM centers organize provincial CDM training programs, but these differ in frequency and scope. Some only carry out essential training in the framework of their projects, involving collaborations between Chinese and foreign partners; others carry out self-financed training in order to establish contact with potential project owners and source projects. CDM centers also publicize CDM handbooks, which complement the training. These handbooks have been similar in their content, and normally include an introduction to climate change, the Kyoto Protocol and the CDM, an overview of the current Chinese CDM institutions, regulations, and registration procedures, and an outline of CDM potential of sectors of the province concerned.
- *CDM project development.* All CDM centers are active in collecting project idea notes (PINs), developing PDDs, and initiating CDM projects. Depending on the donor's approach, programs may be supplemented in a variety of ways: ADB focuses on small-scale projects, Denmark on programmatic CDM projects, UNDP on CDM projects with explicit benefits to the MDGs, and France on projects that promote bilateral cooperation in clean technologies.

The puzzle of wide variation in performance

The idea of establishing provincial CDM centers in China was an innovative one, which might prove valuable in increasing the capabilities of local companies to make use of the CDM. Although India has arrived at a similar approach (see Chapter 8), the Chinese model was established earlier and has been implemented on an astonishing scale, now covering 27 of 31 possible provincial CDM markets. This book asks whether the model is successful, worth supporting, and transferable to other countries, and thus it focuses on assessing the effectiveness of the CDM centers in terms of their three main objectives: raising awareness, capacity development, and project development.

Prior to any empirical work, the centers' performance can be checked preliminarily for their effectiveness in project development, for which data are easily available. It makes sense to compare their success rate in terms of both the numbers of CDM projects developed and the CERs that these projects generate. The numbers of projects give a rough estimate of how successful the centers are in project sourcing and development, also in comparison to other project developers (see Figure 2.9). Because the provinces all vary in their individual CDM potential, one cannot expect all CDM centers to have the same rate of project development regardless of different natural and industrial framework conditions. For this reason, one indicator for centers' performance is the market share of projects that have been developed in a province by the center in comparison to other project developers. Using the market share of projects instead of checking only the absolute number of projects as an indicator does help to control for 'CDM potential,' and for other provincial framework conditions that can influence the ability of all project developers (CDM centers and private consultancies) to develop CDM projects (see Table 2.3).

It makes sense to compare CDM centers in terms of the volume of CERs that are expected from their projects, because CDM projects that generate a large number of CERs tend to be more complicated than projects that generate fewer CERs, although some industrial gas projects

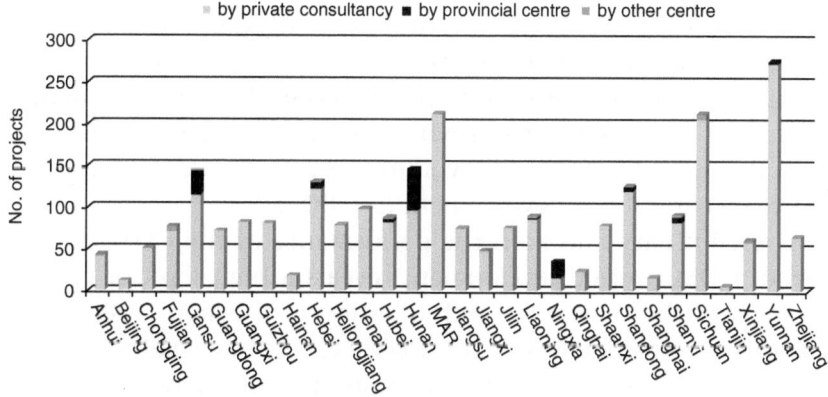

Figure 2.9   CDM projects under validation developed by CDM centers
Source: Based on UNEP Risoe CDM/JI Pipeline Analysis and Database, December 1, 2010.

Table 2.3   Numerical and CER shares of projects developed by selected CDM centers in 2010

Province	Number of CDM projects		Total number of kt CERs until 2012	
	Total	Developed by CDM centres	Total	Developed by CDM centres
Anhui	43	1 (2%)	27,852	114 (0.4%)
Fujian	71	6 (8%)	41,468	11,758 (28%)
Gansu	142	27 (19%)	60,874	6,633 (11%)
Hebei	130	9 (7%)	56,514	6,549 (12%)
Henan	98	1 (1%)	62,390	405 (0.6%)
Hubei	86	5 (6%)	32,566	1,323 (4%)
Hunan	146	50 (34%)	48,083	14,083 (29%)
Inner Mongolia	212	3 (1%)	107,847	910 (0.8%)
Jiangxi	48	1 (2%)	12,616	72 (0.6%)
Liaoning	88	4 (5%)	92,904	1,169 (1%)
Ningxia	36	20 (56%)	12,736	7,900 (62%)
Shandong	124	5 (4%)	134,908	729 (0.5%)
Shanxi	89	2 (2%)	116,546	4,683 (4%)
Sichuan	205	7 (3%)	135,794	6,481 (4%)
Xinjiang	58	3 (5%)	28,765	1,639 (6%)
Yunnan	275	3 (1%)	91,940	597 (0.6%)
Total in China	2,586	151 (6%)	1,710,470	65,045 (4%)

Source: Based on UNDP Risoe CDM pipeline, December 1, 2010.

are relatively easy to develop. The number of CERs generated by a CDM center can therefore be an indicator of its ability to develop complicated, time- and skill-consuming project types. Because the volume of CERs is also dependent on a province's CDM potential, it makes sense to take the market share of CER volumes as an indicator for the centers' performance in project development (see Table 2.3).

This preliminary check on the performance of the CDM centers in terms of CDM project development reveals a high level of variation. This suggests the central research question of this book: How can the variability in effectiveness of the CDM centers be explained? The following chapter discusses how this question can be approached from a theoretical angle, and finally describes how four provincial CDM centers have been selected for detailed analysis.

3
The Role of Agencies in Diffusing the CDM

The problem at hand: lack of CDM diffusion and market development

The problem that led to the CDM center initiative was the lack of appropriate CDM market conditions, which thus became a barrier for local companies in using the CDM for their projects. Barriers hampering diffusion included an absence of awareness, knowledge, capacity, availability of project financing, and political support. The saying that the CDM 'is like a cake falling from the sky,' which was very common in China around 2005 and 2006, demonstrates the general public skepticism towards the CDM. As an institution it was not known or trusted, and potential market participants were not able to access its benefits and risks. Being risk-averse in nature, most local companies thus refrained from taking the CDM into consideration for project financing. Because the market did not evolve quickly and showed early signs of market failure due to imperfect information, the Chinese government decided to support the initiatives of several local government officials to establish provincial CDM centers.

These CDM centers should help to overcome barriers to the CDM market. Innovation theory, which aims to explain product diffusion, should provide guidance on the best means by which to diffuse the CDM and about the role of local agencies as diffusion catalysts. Although innovation theory has not yet been used in the CDM context, its assumptions are transferable because the CDM resembles an 'innovative product,' and the CDM centers are comparable to 'catalyst actors.' Also, explanatory factors used in innovation theory, such as information cascades, learning processes and other group dynamics, and a focus on state actors as catalysts also seem to be explanations worth exploring for CDM diffusion.

Innovation theory: how to speed up the diffusion process

Innovation is a synonym for change. It may be change in terms of new products, but it can also mean the emergence of new processes. Of particular interest here is the way in which CDM-related interventions can change environmental governance – and it therefore makes sense to choose a broad definition of innovation that incorporates not only technologies, but also political processes (Tews, 2005; Holzinger et al., 2007). Schumpeter understood innovations as referring to new products, new processes, new ways to penetrate markets, new ways of supplying goods, or new industries (Schumpeter, 1961 [1912]: 66). Diffusion is classically conceived as a linear process following an S-shaped curve (Schumpeter, 1961 [1912]; Schumpeter, 1939). According to the theory, an innovation is initially diffused on the market only slowly, until it reaches an adoption rate of a about 15 to 20 percent (Golder and Tellis, 1997; Tellis et al., 2002). At that tipping point, an innovation usually 'takes off' and diffuses quickly, until the innovation reaches maturity and the market becomes saturated. Thus, an innovation continues to spread to the remaining late adopters only slowly (Jaffe et al., 2002: 46, Foxon, 2003: 3).

For the purposes of this book, the general models on product diffusion need to be applied in particular to the CDM diffusion process. One fundamental distinguishing feature of the CDM in comparison to other 'products' is that it does not need to go through phases of research, development, and commercialization, because CDM project design was determined by the members to the Kyoto Protocol and cannot be modified by individual market actors. The CDM is, as it were, a given international 'product' for which demand and supply cannot be increased by modifying its design, but only by increasing awareness of its existence, by increasing capacity for judging project eligibility and profitability, and by developing as many successful CDM projects as possible, as showcases for others to imitate.

Innovation theory is instructive about the possible interventions that can speed up product diffusion. Here, a closer look at the micro-level 'innovation-decision' process can help: What makes potential adopters accept the new product, produce it, and diffuse it on the market? And what can be the role of the state in this process of adoption? Why some innovations become more quickly adopted than others depends on a variety of reasons and circumstances. These include, most prominently, information provided to the adopter and the general reduction of uncertainty – as well as learning and imitating processes, the features of the innovation itself, and the achievement of a critical mass of adoption.

Provincial CDM centers can be considered to be diffusion catalysts. Setting up agencies as public or private actors to facilitate diffusion processes has been an approach of market development in other contexts in the past. The approach of setting up agencies that induce change, especially in markets, has been used frequently in a variety of political fields and policy areas. For example, as catalysts for renewable energy markets, agencies can contribute to overcoming an emerging market's shortcomings, such as lack of information, networks, finance and technical expertise. When so-called 'market facilitation organizations' (MFOs) are successful, they help to reduce the risk of engagement for market participants, because the perceived institutional efficiency of a market is one factor influencing an investor's decision as to whether to invest (OECD, 2007a: 21). Using agencies with a clear mandate to promote and diffuse an innovation is a good approach in situations in which diffusion by mere regulation would not be sufficient – for example, because the rule of law and rule enforcement are weak, and business interactions are relationship-focused and less reliant on contracts or laws. A change agent can be defined as an individual or organization that 'influences clients' innovation-decisions in a direction deemed desirable by a change agency' (Rogers, 2003: 400). Agencies that have the mandate to speed up innovation diffusion can be regarded as just such change agents. Also, the provincial CDM centers can become crucial catalysts for initiating and steering local CDM markets. In such cases, the centers themselves are not supposed to bring change to an existing economic-technical regime, but should act only as facilitators to enable other market participants to take up new roles, change their behavior, and ultimately modify the structural conditions in which they operate (see, for example, GTZ, 2000: 9).

Practical experiences with diffusion catalysts

The closest analogy to the work of the CDM centers can be found in regional development economics, in which so-called 'business development service centers' (BDS centers) are employed as one instrument to support markets and private entrepreneurs in less developed regions or countries. In practice, there are many different kinds of market facilitation organization, leading to a bewildering array of labels: 'BDS centers' (OECD, 2007a: 47), 'business support institutions' (UNECE, 2002: 6), 'innovation relay centers' (UNECE, 2002: 9), 'small industry development organizations' (SMIDAs) (Assunção et al., 1993: 195), 'industrial extension services' (OECD, 1995a: 19), 'knowledge-intensive service activities' (KISA) (OECD, 2006a), 'knowledge-intensive business services' (KIBS) (Bilderbeek et al., 1998; Miles et al., 1995; Toivonen, 2004), 'market facilitation organizations' (Martinot, 1998), and 'technology and start-up centers' (Findeis, 2007) are all terms that refer to organizations supporting local economic development, albeit by different means. Their activities can include training and capacity building, technical assistance in productive investment, consultancy and promotional programs, business information, network development, and marketing assistance.

Business development service centers

The organizational structure of BDS centers can take many forms, and is also heavily disputed in the literature. Although BDS centers have traditionally been public organizations, there has been a strong trend since the 1990s towards establishing privately run BDS centers (Meyer, 1992; ILO, 1997: 49). In the recently emerging market-based approach to market development, donors and their partner governments take only a catalyzing role, enabling other agencies to provide the services to the target group. Services are delivered through private sector suppliers, thus 'facilitating the expansion of markets rather than providing services' (Miehlbradt and McVay, 2003: 20; Steel et al., 2000). This shift from a public towards a private setting for BDS centers derives from concerns about ensuring their sustainability and customer focus. In order to ensure financial sustainability, money no longer flows directly into service provision, but is instead provided for establishing framework conditions that encourage private service providers to enter new markets, or to expand their services to under-served markets (Miehlbradt and McVay, 2003: 15; Goldmark, 1996). The shift towards private service companies reveals that the provision of BDS services is seen as 'a tool to [promote] private sector development and not an end in [itself]' (OECD, 1995a: 30). The climax of this paradigm shift was the issuing in 2001 of the Market Development Paradigm by the Committee of Donor Agencies for Small Enterprise Development (CDASED/World Bank, 2001).

Programs for BDS centers usually aim to establish centers with the following institutional characteristics:

- *Organizational capacity.* BDS centers should be run in a business-like manner, with clear corporate culture, client group, and performance measurement.
- *Managerial capacity.* BDS centers should be organized in a decentralized manner, should build on participation and ownership, and should show high autonomy from the state. Good managerial

and technical skills and capabilities, entrepreneurial attitudes, and human capital are all important.

- *Financial capacity.* BDS centers should strive to become financially independent from their government as well as their donor programs, securing financial sustainability and independence. Means to this end include charging fees for services, economies of scale, services provided to groups rather than individual clients, and diversified sources of funding.
- *Technical capacity.* BDS centers should have good human resources departments with the ability to specialize in areas of expertise, and to meet changing client demands through service and product innovation.
- *Outreach.* The density of BDS centers' regional presence is important for their reputation and for the firms' responsiveness. A demand-led approach, with proximity to clients and targeting those with the greatest demand for BDS services, is important. Embeddedness in the local business environment is helpful in reaching out to potential customers.

(UNECE, 2002: 15f, Miehlbradt and
McVay, 2003: 47, OECD, 2007a: 42)

The core function of BDS centers is to help private actors by reducing uncertainty and risk for private investments. BDS centers can reduce the range of transaction costs either by directly providing services that help customers to overcome market problems or by indirectly matching clients' demands with suppliers' services. Concerning the timing of interventions, traditional market development programs became active at the delivery stage, most often directly supplying services. Because this direct intervention can distort the market, many authors now suggest intervening at the pre- and post-delivery stages (Miehlbradt and McVay, 2003: 56). Due to its potential power to distort the market, the golden rule of market facilitation is that public interventions should only be as strong as is required to address the market failure under consideration (Miehlbradt and McVay, 2003: 56).

There is a tendency to see BDS centers as transforming from direct service deliverers into catalysts between customers and private service suppliers. Instead of activities that have been decided upon a priori by the funding government or donor, BDS centers are encouraged to set up their individual and flexible support strategies based on customers' demand. Flexibility, customer-orientation, and adopting the role of catalyst have all proved decisive for the effectiveness of BDS centers.

DNAs as CDM-promotion agencies

The CDM market is also one example of a specific sector for which organizations have been established with the purpose of market development. The most obvious support institutions are the 'designated national authorities' (DNAs). Although their main function is to act as a representative institution of the host-country government that can give official approval to CDM projects, DNAs can also serve the function of promoting and supporting markets. A large literature is available on experience with DNA establishment, and can be drawn upon for CDM institution-building at the local level.

A DNA should ideally be a 'one-stop shop' for public and private actors interested in implementing CDM projects. CDM-related institutions should be able to a) possess adequate information on CDM modalities and procedures, and b) facilitate CDM transactions in a timely and transparent manner (Ellis and Kamel, 2007: 26). An ideal DNA should be:

- *responsive* – able to respond in a timely manner to inquiries and to make decisions promptly;
- *flexible* in its operational procedures, so it can adapt quickly to new developments;
- *sustainable* in terms of its ability to ensure long-term financial funding;
- *efficient* in terms of the relationship between budget and staff size to the organization's activities; and
- *transparent* in its operations and procedures.

(Ellis and Kamel, 2007: 28)

DNA functions include establishing national guidelines for project development and approval, especially with regard to national criteria on sustainable development. Once institutional rules are set, the main function of the DNA is the actual approval of CDM projects – for example, for their sustainable development benefits and their voluntary nature. Besides these formal requirements, DNAs often act as promoters of their country's potential for CDM projects, and engage in match-making activities with foreign investors for CDM projects by, for example, providing project pipelines on their web-pages, or hosting international conferences. DNAs often also adopt the role of facilitating national CDM markets, in which they use either a governmentally driven approach, where only the DNA is responsible for information sharing; or a private-public partnership approach, in which the DNA provides data for baseline development but private entities conduct experience-sharing events; or a purely privately driven approach for CDM service provision (Babu and Michaelowa, 2003: 23). But caution is necessary when conflicts of interests arise – for example, when the promotional activities of a DNA interfere with its neutrality in project approval.

Unlike BDS centers, the usual approach towards CDM promotion in CDM host countries is to set up DNAs that are part of or attached to the government. This makes sense, since the core function of DNAs is to confer the country's approval on CDM projects. (Though one might also consider a private certification company providing this function, using to assessment criteria defined by the government. In reality, technical project approval is often already done by external researchers rather than by government officials). These public DNAs mainly provide services directly, instead of linking customers with an external provider. The advantage of the public character of a DNA as a CDM promoter lies in its assumed ease of access to consolidated information, and in its high political standing. Based on these considerations, ideal public DNAs would be 'one-stop shops' that, by virtue of their public setting, are able to provide transparent and reliable services and consolidated information.

Lessons learned from practical experiences

No canon of ideal best practices, or easy recipes to copy, has yet emerged from the experiences of market facilitation organizations in relation to how these organizations can be most effective in their work (Ellis and Kamel, 2007: 42). Nevertheless, some useful factors contributing to effectiveness may be identified. There is a consensus in the literature that market facilitation organizations should focus on their role as catalysts instead of directly delivering services (Miehlbradt and McVay, 2003: 14). Still contested, however, is what type of organization is most suitable for providing market facilitation services. These might be private sector providers, local or international NGOs, consulting firms, governments, and funders – all of which are seen to have both advantages and disadvantages in their ability to facilitate markets (Miehlbradt and McVay, 2003: 48). In general terms, there is a common understanding that market facilitation organizations are successful if they are able to reduce the risk of engagement for private market participants. Once these risks have been limited, private actors can become drivers of further market progress.

From the empirical approaches to market facilitation organizations examined in this section, it is clear that there is a general tendency for them to be established as private companies providing services

Table 3.1 Factors determining the effectiveness of market facilitation organizations

	Organizational strength	Intervention strategies	Additional factors
BDS centers	Tendency towards private setting	Mainly indirect service provision	Customer-orientation, flexibility, catalyst role
DNAs	Public setting due to Kyoto Protocol requirements	Direct and indirect service provision	One-stop shop, reliability

indirectly. Practical experience also underlines the catalyzing character of such organizations and their match-making activities as crucial determinants of success (see Table 3.1). In addition to the institutional requirements of having an adequate incentive structure and resources, a review of practical experience once again corroborates the need for good management practices and organizational sustainability.

Analyzing CDM centers' effectiveness in CDM diffusion

The empirical part of this book aims to analyze the effectiveness of provincial CDM centers as one sample for hybrid actors' performance in local climate governance in China. The following section provides a review of factors derived from innovation theory and the practical experiences summarized above, which might be relevant in explaining the effectiveness of Chinese provincial CDM centers.

Organizational strength

The organizational strength of provincial CDM centers and of private consultancies is one important factor in analyzing their effective impact on market development. Although the agencies themselves are only an indirect means to reach policy objectives, their strength is a crucial factor in influencing their intervention strategies (see OECD, 1995c: 5). If agencies are weak, they can fail despite having the best strategies; if agencies are strong, even weak strategies may have some effect. Organizational strength depends on several internal and external factors. Internal factors can include:

- *Incentive structure.* In order to have a high degree of motivation, an organization should have a clear mandate and adequate autonomy in pursuing its goals. This would require the Chinese CDM centers to have clear mandates governing their responsibilities. Practical experience has shown that a center should take on the role of catalyst or change agent rather than trying to provide services directly. In order to live up to its mandate, a center requires adequate space for its own maneuvers, and should ideally have full autonomy from the government, independence from donors, and a performance-based remuneration system.

- *Public or private status.* The choice between public and private status has a relevance to innovation theory, which considers the right timing and degree of public and private actors' activities as crucial for diffusion processes. Public actors are thought to be required for kick-starting markets, while private actors are usually more efficient in providing tailor-made, flexible services. Empirically, there is a clear preponderance of private market-facilitating organizations.
- *Resources.* Resources are the prerequisites for developing and implementing intervention strategies. They can include financial resources and human resources, but also information resources –all determine the ability of the organization to gain access to important information. Human resources can be measured by the functional skills and local knowledge of staff, and by staff motivation. Financial resources can include the consulting fees paid by clients, technical assistance grants from donors, and Chinese central or provincial government resources, or – in CDM markets – negotiated future fees from the sale of resulting CERs. Information resources are determined by the ability of the organization to gain access to important information and manage it properly (OECD, 1995c: 28).
- *Sustainability.* If an organization is able to adapt to changing circumstances, it can ensure its existence into the future. After initial start-up subsidies from international donors, these agencies should be able to recover costs and achieve financial sustainability (OECD, 2005d: 30).

External factors can include:

- *Embeddedness.* Strong embeddedness of an agency should ease cooperation. Horizontally, it can include good relations and networks with governmental actors, businesses, and private organizations. Vertically, these relations and networks can operate at various levels, which can include – depending on the location of the agency itself – local, regional, national, or international actors.
- *Political standing* is established through support for the agency by political actors, foreign government agencies, or foreign NGOs, all of

which increase its resource availability and can help with knowledge and technology transfer (Willems and Baumert, 2003: 13).

Most of the literature on institution-building for the environment or for private-sector development includes similar checklists of what are perceived as important characteristics of strong institutions (see for example UN, 1982). But it should be clear that institutional capacity is more than just the sum of its parts. Instead, organizational strength refers to the capability of organizations in using their skills and structures, not just to the availability of such skills (Fukuda-Parr et al., 2002: 10).

Intervention strategies

As we have seen, several internal and external factors shape the organizational strength of agencies such as the CDM centers. While these factors constitute the capacities based on which an agency can operate, its effectiveness depends mainly on the correct choice, design, and implementation of intervention strategies that make use of these capacities (Jänicke and Weidner, 2002). Effective interventions for market development depend on a variety of factors. Intervention strategies can be differentiated according to their single- or multi-purpose character, their timing, and their outreach; they can vary in number, variety, and target specificity of units reached. One fundamental insight can be applied to all intervention strategies: wherever possible, CDM centers should take up the role of catalyst, facilitating market development rather than being a direct provider of services.

No blueprint approach yet exists for an agency's intervention strategies for speeding up diffusion processes. But good intervention strategies offer a match between the agency's mandate, its capacities, and its external framework conditions. Based on the literature review of innovation theory and on the mandates of the CDM centers to facilitate their local CDM markets, the following paragraphs suggest some possible strategies. These focus on a) information dissemination, b) creating learning opportunities, and c) facilitating CDM projects – three areas of intervention that have already been identified as the core fields of activity for CDM centers (see Chapter 2).

Information dissemination

Many practical cases have shown that innovation diffusion rarely occurs at an optimal speed. There are two types of situation in which diffusion takes place more slowly than might be expected. First, innovations often do not diffuse at all, or do so painstakingly slowly, despite their social or environmental benefits and the political will to push them forward (see examples of slow diffusion of energy-efficiency standards in the building industry in Jaffe and Stavins, 1995). In such situations, adoption costs seem to be too high for the potential adopters, despite the social and political benefits of the innovation. Second, some innovations fail to diffuse sufficiently despite their low costs. This phenomenon has been dubbed the 'paradox of high potential but low adoption' (Shama, 1983). It frequently occurs in energy markets and has been observed, for example, in cases of innovations of increased energy efficiency that are very cost-effective to adopt but which nevertheless diffuse slowly on the market because potential users are locked into existing technological preferences (this situation is also termed an 'energy-efficiency gap' – Golove and Eto, 1996: 5). In both cases, information plays a crucial role in overcoming adoption barriers.

Uncertainty about an innovation usually exists throughout all stages of diffusion (Barreto, 2003: 5). The question of how to deal with uncertainty is a trade-off for each adopter: either one accepts uncertainty and the risks it incorporates, with the possibility of reaping the advantages of being an early adopter, or one takes a risk-averse position and waits until others have had their experiences with the new product, and have thereby reduced the uncertainty associated with it. Uncertainty might easily be overcome by the gathering of sufficient information about the innovation and its present and future market prospects, because information about risk renders it more manageable.

Information about a new product is often not sufficiently available to the public. In most instances, information has the characteristics of a public good, because market actors cannot be excluded from its usage and there is no rivalry in using publicly available information (Jaffe et al., 2002: 48). In some instances, however, information can be a private good – belonging, for example, to an intelligence service or market competitor (Keck, 1987: 141). Information about the experiences of early adopters of innovations can also become a positive externality if it eases the adoption process of other companies that subsequently adopt the same innovation, thereby profiting from the experiences of early adopters. Information asymmetries can occur if such information is spread unevenly among market participants. Imperfect information is even seen as 'endemic' to the process of technological change, and its markets for information are seen as 'notorious[ly] ... imperfect' (Arrow, 1962; Stoneman and Diederen, 1994: 920). Information deficiencies can take the form of inaccurate or insufficient information about the existence, the features, and the future outlook of the innovation. Thus, due to it being a public good and thus an externality, information is likely to be provided by the market at a socially suboptimal rate, constituting the phenomenon of market failure (Akerlof, 1970).

Market failures like information deficiencies and asymmetries can slow or even freeze the diffusion process (analyzed for energy markets in Fisher and Rothkopf, 1989). There are several sources of market failure that relate to innovations. In addition to a lack of information, these market failures can include a lack of enforceable property rights, distorted market power, principal-agent problems, constrained capital financing, and positive or negative externalities (Stoneman and Diederen, 1994: 920; Jaffe et al., 2002: 48). The consequence of market failures is often 'wait and see' behavior on the part of potential adopters, because these act in a risk-averse manner, watching their peers' behavior first and tending to postpone adoption (Pindyck, 1991; Hassett and Metcalf, 1996).

Uncertainty, imperfect information, and the bounded rationality of actors (their limited capacity to take all available information into account in their decision-making – see Simon, 1972) can also explain why CDM adoption was slow in the early carbon market in China. The early CDM market in China reflected the 'high potential but low adoption' paradox because the CDM would have been very useful for many potential adopters, even from a rational-choice perspective, but was actually used by only a very few companies. As was noted in Chapter 2, the diffusion of the CDM faced several barriers in China. At the provincial level, these barriers were mainly a lack of awareness and knowledge of the CDM and eligible projects, a lack of capacity for CDM project development, and a lack of political and policy support. In the language of innovation theory, factors such as imperfect information, uncertainty, constrained capital financing, and some positive externalities are all important in explaining the slow takeoff of the Chinese CDM market in 2005.

Public interventions for information provision are generally welcomed as a means to reduce transaction costs – especially search costs – for private business actors (Stoneman and Diederen, 1994: 920, Morgenstern and Al-Jurf, 1999). Public information provision is especially welcomed in paradoxical situations in which innovations appear cost-effective, but are not yet widely utilized because the market fails to provide adequate information (Anderson and Newell, 2004). The economic rationale of public interventions resides primarily in providing the public goods of knowledge and information.

Information programs are thus one popular policy to speed up diffusion processes. They aim to increase awareness of the innovation and its merits, and they often also offer technical assistance with their implementation. These programs employ various strategies – for example, media campaigns, advertising, and product labeling. In addition to alerting firms to the potential benefits of an innovation, these programs often provide accurate information about the innovation's features, and thereby reduce the uncertainty and risk associated with adopting them (Newell et al., 2006: 569).

But exactly what kind of information makes companies accept an innovation? Research on the acceptance criteria for new innovations emphasizes the importance of information being targeted specifically at those who need it. Because potential adopters differ in their individual characteristics (for example risk-averseness), information messages are most efficient if their content is adapted to their target group. It can be an important strategy of a CDM center to prioritize the order in which it disseminates information to various target groups. If potential opinion leaders can be persuaded early on about the advantages of the CDM, their positive example will pull others to adopt it more quickly, thus speeding up the innovation process. In the Chinese cases, one imperative for the CDM centers should be initially to target opinion leaders such as leading government officials and high-status SOEs, or private enterprises.

Another factor that speeds up diffusion is the framing of the innovation: the type of information provided should cater to the addressees' needs. Innovations tend to be accepted more widely and thus diffuse more quickly if they are presented as having the following characteristics.

1. *Relative advantage* is the degree to which an innovation is perceived as better than the existing product or service.
2. *Compatibility* is the degree to which an innovation is perceived as consistent with existing values, past experiences, and needs of potential adopters.
3. *Complexity* is the degree to which an innovation is perceived as difficult to understand and use. New ideas that are simple to understand are adopted more rapidly than innovations that require the adopter to develop new skills and knowledge.
4. *Trialability* is the degree to which an innovation may be experimented with on a limited basis. New ideas that can be tried on a small scale and for a limited period will generally be adopted more quickly than innovations that are not divisible.
5. *Observability* is the degree to which the results of an innovation are visible to others. The easier it is for individuals to see the results of

an innovation, the more likely they are to adopt it. Such visibility stimulates peer discussion of a new idea.

(Tornatzky and Klein, 1982;
Shama, 1983: 156; Rogers, 2003: 15)

This would imply that the CDM should be understood as relatively advantageous in comparison with other financial mechanisms, being compatible with existing requirements for energy-related projects, easy to understand, and easy to use.

Since the diffusion of innovations is basically a means of bringing change to a social system, different channels of communication can serve as important devices to reach different social actors. Depending on the phase of the innovation-decision process, different communication channels are appropriate. In the early phases, mass media channels can spread information about the existence and the rough features of an innovation most rapidly and efficiently among potential adopters (Shama, 1983: 159). For directly persuading an adopter to take up the innovation, however, experiences from innovation research reveal that interpersonal communication is more effective in persuading an individual to accept an innovation (Rogers, 2003: 18). The phenomenon of 'information cascades' rapidly spurs diffusion if a majority – or at least an prominent cluster – of peers adopts the innovation. Networks can become catalysts for learning processes, because they encourage learning from peers, enable experience-sharing, and stimulate compatibility of the innovation with the needs of its adopters (see Silverberg, 1991). One crucial component of any information-dissemination strategy is therefore the use of appropriate channels of communication. Innovation theory and experiences with DNAs have shown the importance of using popular channels for information dissemination. Three strategies exist, and they can complement each other. The first strategy is to use mass media channels such as TV, radio, print media, and the internet to reach a broad and general audience. For example, one popular approach for information dissemination is the establishment of a website (Gupta and Michaelowa 2005). The second strategy is to use peer networks of the potential adopters, and to make opinion leaders the advocates of the CDM. The third strategy is to use interpersonal channels such as workshops, excursions, exchange visits, joint conferences, and person-to-person or group training sessions, all of which involve face-to-face exchanges between two or more individuals.

Capacity development

Learning is another major factor explaining the diffusion of innovations and of technical change in general (Wright 1936; Arrow 1963;

Smith, 2001: 8). The acquisition of knowhow is also referred to as 'competence building' (Edquist, 2005: 192). The rationale for enabling learning opportunities is based on so-called 'learning curves' that have been observed for many diffusion processes, and which imply a positive feedback between learning and diffusion: the more a product becomes diffused, the more information is available about its advantages and disadvantages, creating knowledge 'spillovers' to other potential adopters, who might be persuaded by the positive experiences of their peers to adopt the product themselves, thereby further diffusing it on the market (Stokey, 1986; Argote and Epple, 1990; IEA, 2000; McDonald and Schrattenholzer, 2002). The advocates of an innovation frequently offer free training and other help to potential users in order to bring down their costs of adoption (Shapiro and Varian, 1999).

Capacity development programs are thus another popular means of facilitating diffusion processes. They aim to enable the potential adopters to understand an innovation, evaluate its costs and benefits, and eventually persuade them to try using it. Capacity development programs often entail activities such as workshops, training sessions, excursions, and experience-exchange programs.

In economics, the concept of 'learning-by-doing' is part of the approach that regards markets as being in a continuous process of becoming, resembling an 'enacted evolutionary structure' (Arrow, 1962; Snehota, 2004: 26). Innovation theory stresses the importance of learning-by-doing for the diffusion of innovations. In contrast to the passive intake of information, as in traditional classroom teaching, learning-by-doing emphasizes the conversion of knowledge into knowhow by practical experience. 'Learning-by-doing' is a process in which individuals continuously revise their interpretation of a situation through the experience and understanding they gain from applying that knowledge (Arrow, 1962: 155). In order to move from hearing information for the first time to being able to use it effectively, the learner needs to use and experience the knowledge (Dosi, 1988: 223). Learning-by-doing is therefore not a one-off endeavor, but a continuous process. It implies continual improvement. Strategies that aim for learning-by-doing should therefore offer training that includes hands-on experience of the innovation, repeated over time and adjusted to the new knowledge levels of learners.

'Learning-by-interaction' is another important mechanism in the adoption process and refers to the observation and sharing of the adoption experiences of companies that already use an innovation (Rogers, 2003: 18, Lundvall, 1992; Christensen and Lundvall, 2004). Ideally, such companies would be opinion leaders. In innovation theory, 'opinion leaders' refers to individuals 'in a unique and influential position in their system's communication structure,' and who have technical competence, are socially accessible, and show conformity to the system's norms (Rogers, 2003: 27). By sharing these experiences and providing a subjective assessment of the pros and cons of an innovation, the early adopter creates a 'positive externality' for others by generating information about the existence, characteristics, and success of the new technology (Jaffe and Lerner, 2004: 167). If model solutions exist, the attractiveness of the innovation and its speed of diffusion are likely to increase (Tews, 2005: 267). Strategies that aim for learning-by-interaction should therefore offer opportunities to meet with peers who have already adopted the innovation, and are willing to share their experiences and advise other potential adopters on the process of adoption. Classroom teaching should therefore be complemented by exchange visits and excursions to companies already using the innovation.

The argument that a learning curve exists within the CDM market, and that learning effects would lead to a reduction in the transaction costs of CDM projects, was advanced very early on in the debate (Michaelowa, 1996; Puhl, 1998). Transaction costs were seen as one of the major barriers to CDM adoption (Michaelowa and Stronzik, 2002: 11f). It was therefore generally accepted that a learning curve would make the later phases of implementation less resource-intensive, because project owners would have learned where to expect high transaction costs (PriceWaterhouseCoopers, 2000). Calculations conducted by Michaelowa led to the assumption that annual implementation costs (of monitoring, for example) would account for only 80 percent of the first year's costs in the subsequent years of the project's lifetime (Michaelowa and Stronzik, 2002: 16). Future research is still required in order to assess which kinds of transaction costs are likely to be reduced by such a learning curve and which are not (Del Río, 2007: 1,363).

In order to make use of the learning curve and to persuade potential adopters to make use of the CDM, one fundamental task of the provincial CDM centers is for them to enable a learning process and build CDM capacity among potential project owners regarding their CDM options. Training, for example, on CDM requirements, project eligibility, and CER trading should enable potential adopters first to make cost-benefit calculations and assessments of whether the CDM is worth considering for their individual project. Besides such a rational-choice approach to learning about the advantages and disadvantages of an innovation, the evolutionary approach, with its emphasis on learning-by-doing and learning-by-interaction, should also be considered. In addition to traditional training, there might be learning opportunities that include showcase demonstrations of innovations, and to meet and interact with companies that already use the CDM. Because a learning curve can also be assumed for each individual user, learning opportunities should ideally be tailor-made to the needs of the client, taking into account his or her increasing knowledge levels.

Facilitation of CDM adoption by successful CDM project development

Although potential adopters may have enough information about the CDM, as well as knowing whether their project is CDM eligible and having a general interest in using the CDM, there are often still many barriers that prevent their ultimate adoption of the innovation. These business-related barriers usually decline with the learning curve and diffusion rate of the innovation, because adequate experience of how to use an innovation has been gained, spread, and shared between potential adopters. Thus, the most direct way for CDM centers to support CDM diffusion is to develop successful CDM projects themselves that demonstrate that adopting the CDM is a good decision. One way of encouraging potential project owners to adopt the CDM is to share positive experiences with them about projects that have already been successfully developed.

The effectiveness of CDM centers in CDM project development is likely to be based on their ability to source projects and on their own expertise in CDM project requirements. But experience with BDS centers has revealed that direct diffusion measures pose the risk of market distortions by government- or donor-funded market-facilitation agencies like the CDM centers. These experiences suggest an indirect, catalyzying role for these agencies that aims to establish the enabling conditions for private business actors to provide CDM project-development services.

In line with this approach, the CDM centers should only facilitate the CDM adoption process by assisting companies in finding suitable project-developing consultancies, financing institutions, and potential buyers. CDM centers can facilitate CDM adoption by encouraging potential project owners and potential CER buyers to meet, communicate, and come to an agreement resulting in an ERPA contract. In addition, the CDM center can ensure political support of the CDM in the form of provincial-level CDM policies that are supportive and stable, providing foreseeable framework conditions, and are effectively enforced. The availability of such political support is likely to strengthen the willingness of potential adopters to use the CDM.

Intervention is helpful in kick-starting product diffusion, but the ultimate decision as to whether a product is adopted and used on a large scale depends on whether it is able to reach the 'tipping point' in the diffusion process. This tipping point is reached when between 15 and 20 percent of potential users have adopted the innovation. At this point, innovation theory assumes that full diffusion will take place more or less automatically. The ultimate goal of any market-development intervention must therefore be to ensure that the innovation reaches this tipping point, thereby eventually making public interventions to 'pull' the market unnecessary by switching to the 'market push' mode of diffusion. Diffusion usually begins in the niche markets, where the technology may be attractive due to specific advantages or particular applications. In niche markets, valuable experience can be accumulated so that performance improvements and cost reductions may result. Early niche markets help firms to gain feedback from the experience of early users, and to demonstrate feasibility to all other potential users, manufacturers, and other interested actors such as policy-makers (Kemp, 1997). But successful commercialization will depend on further performance improvements and cost reductions ensuring competitiveness (Grübler et al., 1999). If successful, the technology may extend to other markets, and may eventually be able to diffuse within all market segments.

Pulling markets towards this tipping point of quasi-automatic diffusion inevitably involves facilitating adoption by individuals. The later a company decides to adopt an innovation, the fewer will be the remaining barriers to overcome. In addition, profit margins might be smaller due to heavy competition. For each adopter, it is thus of paramount importance to reach a point of equilibrium for adoption costs and benefits. The strategies of information dissemination and capacity building mentioned above, however, contribute only indirectly to overcoming adoption barriers. A more direct facilitation of product usage is often necessary to ensure that a critical mass of adopters is reached.

Even if information dissemination and capacity development strategies have changed the adopters' attitudes and made them willing to try out the innovation, several barriers related to the actual adoption process must still be overcome. These barriers can include a lack of business contacts and experience with the innovation, and a lack of political support for the innovation and its adopters. Intervention strategies that directly aim at diffusing products should therefore contribute to overcoming direct barriers of product usage.

Alternative explanations for local CDM diffusion

Private consultancies that offer their services for CDM project development are direct competitors of provincial CDM centers. Their presence and marketing strategies should have a strong influence on the effectiveness of the CDM centers, at least in terms of CDM project development.

Another important factor accounting for differences in the present performance of the CDM centers, however, is time. Although 27 CDM centers have been planned, only around a dozen have commenced operations. There are even several provinces that have not yet identified their CDM centers. Out of the selected CDM provinces examined in the case studies – the CDM center in Ningxia – was established earliest, and the CDM center in Yunnan most recently. It remains a question for empirical analysis whether starting early on the market brings advantages, or whether starting in a situation of high uncertainty and limited CDM experience implies disadvantages for a center's effectiveness.

Analytical framework

The comparative case studies focus on the question of how to explain the variance in effectiveness of four CDM centers. Effectiveness refers to the degree to which goals are reached. These include chiefly CDM promotion, CDM capacity development, and CDM project development, thereby contributing to the overall mandate of CDM diffusion and CDM market development. Factors that need to be checked for their relevance in explaining the centers' effectiveness include the centers' intervention strategies, which in turn include information dissemination, the creation of learning opportunities, and CDM project development services (see Figure 3.1 for possible causal relationships). Other factors that should be checked for their influence on setting the corridor for CDM activities include variables such as provincial economic and political framework conditions, the global CDM market, and the influence of other private actors besides the CDM centers, all of which might have an influence on CDM diffusion and market development.

A CDM center's success should have a variety of impacts: direct impacts on CDM diffusion in the province concerned, indirect impacts on CDM market development in the province, and very indirect impacts

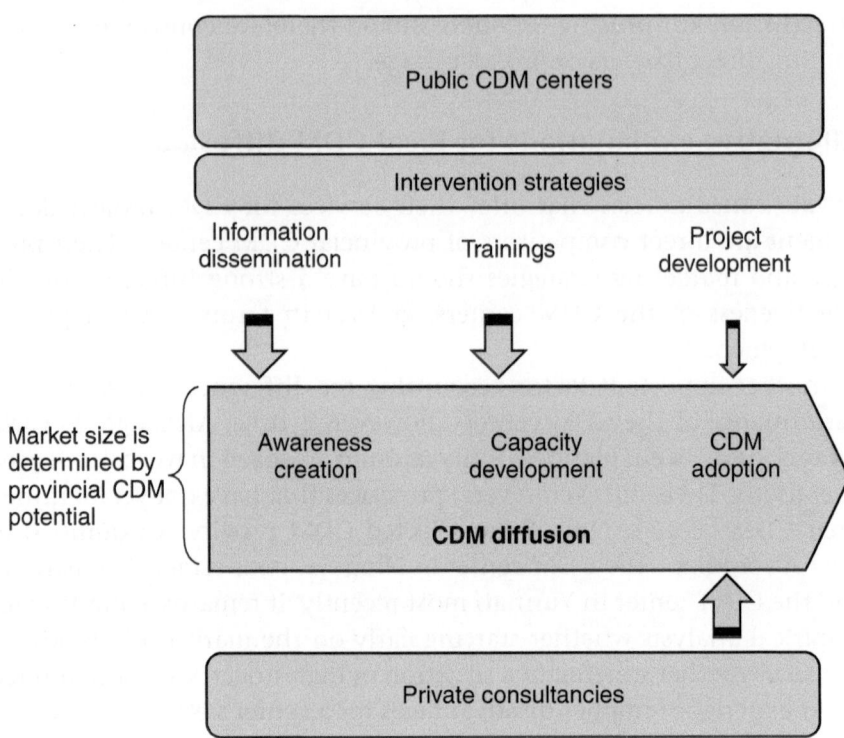

Figure 3.1   Possible factors influencing CDM centers' effectiveness

on the general objective of CDM projects – namely the achievement of GHG emission reductions and sustainable development benefits. On the assumption that CDM projects that are registered, monitored, and verified for their emission reductions will indeed contribute to global GHG emission reductions and sustainable development, these aggregated impacts of the CDM centers are not discussed, as they are only very remotely attributable to the CDM centers.

4

A Case Study on the Performance of Four CDM Centers

Introduction to case studies

This chapter summarizes four case studies on CDM centers and their impacts on CDM diffusion and market development in four Chinese provinces: Ningxia Autonomous Region, Gansu, Hunan, and Yunnan. The effectiveness of the centers relates to several kinds of activities, of which three main types have been selected for detailed analysis. These are a) CDM promotion, b) CDM capacity development, and c) facilitation of CDM adoption. Because it was impossible to know the value of these outputs prior to empirical research, a proxy for centers' effectiveness was used – namely, the number of CDM projects developed by each of the centers versus the total number of projects developed in a province. (Table 4.1 provides data from October 2007, which was prior to the empirical research conducted for this book.) This proxy was selected based on the assumption that CDM promotion and capacity development are prerequisites for centers to be effective in CDM project development. The detailed variance for the centers' effectiveness could only be assessed after the field visits.

The empirical data for the case study are based mainly on secondary literature and on interviews conducted in China in April 2007 and between September 2007 and February 2008. In addition, telephone and personal interviews were conducted with non-Chinese market actors. The 74 qualitative interviews were conducted with representatives of the following stakeholders: CDM centers, provincial government officials (either from the Development and Reform Commission, the Department of Science and Technology, or sometimes from the Environment Protection Bureaus), project owners, project developers, buyers, and Designated Operational Entities (DOEs). Sometimes,

Table 4.1 Development of CDM projects by CDM centers in October 2007

Province	Total number of CDM projects	Number of projects developed by CDM center	Percentage
Ningxia	9	8	88 percent
Hunan	62	27	41 percent
Gansu	57	5	8 percent
Yunnan	80	1	1 percent

Source: Based on UNEP Risoe Center CDM/JI pipeline, October 2007.

representatives of local non-governmental organizations (NGOs) or research institutions were also interviewed for their expert perspective on local CDM market development. All interview partners were granted anonymity, so that only their interview ID and not their names are given as references.

Provincial framework conditions

All four provinces belong to the western and central parts of China, and are underdeveloped in terms of industry and human resources. They therefore share very similar framework conditions compared to China's eastern provinces. The political, economic, and natural resource conditions of a province provide a 'corridor' in which market facilitation organizations can operate and define a limit on what is achievable. Ningxia, Gansu, and parts of Yunnan belong to the Himalayan plateau, while Hunan is situated in the middle reaches of the Yellow River. All four provinces are mountainous and land-locked, and home to many ethnic minorities. Weather conditions are harsh, especially in Ningxia and Gansu, which often suffer from droughts and water shortages.

Political framework conditions are relevant when considering the effectiveness of the CDM centers because CDM promotion and capacity development also depend on the way the government supports the CDM and grants private profit- and non-profit actors access to information, approval, and other channels of involvement (on the importance of local government structures for capacity development, see OECD, 1995b). Although Ningxia has slightly more independence from the central government than the other three provinces, the political framework conditions of the four provinces are very similar.

In all four provinces the secondary and tertiary sectors remain underdeveloped; they have a comparatively low GDP, and thus share similar economic conditions. Each of the provinces hosts at least one of the 18 Chinese regions that are classified by the State Council as 'very poor': for example, Xihaigu region in Ningxia, Dingxi arid land and Northern Shaanxi in Gansu, Wuling mountain region and Jingang mountain region in Hunan, and Hengduan mountain region and the south-western part of Yunnan (Wang, 2001: 93). Ningxia is one of China's poorest regions: average per capita income in the cities increased by 6.5 percent in the five years to 2006, reaching 11,100 CNY, and by 7.8 percent, reaching 3,250 CNY, in the rural areas (Ningxia DRC, 2006). Average annual income per capita was 5,440 CNY in 2003 (Ningxia Science and Technology Department, 2006: 93). Gansu is slightly better off: its GDP reached 192.814 billion CNY in 2006, increasing by 11.7 percent over the previous year. Hunan is the most economically advanced province of the four: its annual GDP reached 914.5 billion CNY in 2007, and its average annual per capita income was 14,405 CNY in 2007, reflecting annual growth of 14.2 percent (National Bureau of Statistics, 2008).

The CDM centers' scope of action is limited by a province's CDM potential, which is determined by its natural resources and its industrial structure. Small geographical regions like Ningxia can be considered to have a well-developed CDM market even if they have only a small number of CDM projects, either because they are small in size or do not have good CDM potential. The four provinces chosen for case studies here generally have good CDM potential, thanks to their natural resources and industrial structure. Due to their often very inefficient industry, these provinces have a good potential for both energy-efficiency and waste-heat recovery projects. Ningxia, for example, was the most energy-inefficient Chinese province in 2006, and consumed 4.5 times more energy in its production processes than the Chinese average (Ningxia Science and Technology Department, 2006: 102). As especially Ningxia and Gansu are coal-rich provinces, they also offer good potential for coal-bed methane projects.

All four provinces also have good CDM potential in the area of renewable energy. Yunnan and Hunan are two of the provinces in China that have very good hydropower resources. Ningxia and Gansu have good potential for solar and wind-power projects. For rural electrification, all four provinces have good potential for the use of biogas and biomass energy projects.

Despite their similar framework conditions, the four provinces show different development pathways in their local CDM markets. Due to its economic underdevelopment and the lobbying activities of its CDM center's director, Ningxia was selected to be one of China's CDM pilot provinces. Although the market itself is very small, making up only 1.4 percent of China's total number of CDM projects and 0.7 percent of the total CER volume in the pipeline in December 2010, Ningxia's first CDM project was ready for validation in mid-2005 (UNEP Risoe, 2010). Gansu's CDM market also experienced an early start, and developed steadily into a middle-sized market from 2005. The CDM was lagging behind in Hunan and Yunnan, but both provinces experienced a sudden rise in CDM project development in the latter months of 2006, and have at the time of writing overtaken Ningxia and Gansu in terms of the number of CDM projects developed. The CDM markets in Gansu, Hunan, and Yunnan are so far dominated by hydropower projects, while Ningxia is also home to many wind-power projects.

The influence of donor programs on a province's CDM market

The first foreign donor project that focused on capacity development at the provincial level was the Canadian program, which supported the establishment of China's first provincial-level CDM center in the Ningxia Autonomous Region. Ningxia was thus the first province to host a provincial CDM center, as early as October 2003. Due to the success of the project, Canada expanded it to incorporate four other provinces, and other Annex I countries swiftly followed in picking up provincial-level capacity development initiatives in China (see Table 4.2 and Map 4.1).

Because the Ministry of Science and Technology (MOST) is the implementing agency for all foreign donor programs for the establishment of provincial CDM centers, these programs are comparable in their influence on the effectiveness of the CDM centers. They differ, however, in their timing, which needs to be considered as a possible additional variable influencing the centers' effectiveness (see Figure 4.1).

Other than in their timing, the donor programs are very similar in their components. They provide initial CDM training for the CDM centers' staff, host provincial-level CDM conferences to lobby for the support of local political leaders, and assist the centers with linkages to potential buyers (often from the donor country) and by providing some financial seed money.

The Sino-Canadian program operated in five provinces: Gansu, Ningxia, Hunan, Shandong, and Hebei. The program's main focus was on PDD development: each province was required to develop at least one PDD. In return, the program provided financial support for the establishment of the CDM center, and Canadian partners introduced potential Canadian buyers to the center (Interview 50). However,

Table 4.2 Overview of selected provincial capacity-development programs

Donor	Location	Time	Objective	Activities
Sino-Canadian Cooperation Pilot Project on Local CDM Capacity Building	Ningxia	2003–05	Exploration of twelve CDM projects and development of three PDDs	Financial support for Ningxia CDM center, training for local project owners and experts
ADB: Opportunities for the CDM in the Energy Sector	Gansu, Guangxi	2004–05	Small-scale CDM project development	Assessment of potential for small-scale CDM projects and training
WB – China's Provincial Climate Change Program	Hubei, Jilin, Shaanxi, and Yunnan	2005–07	Development of provincial-level Climate Change Programs	Training and workshops supporting provincial government in planning process
China-France CDM Capacity Building Cooperation Program	Guangxi, Sichuan, Guizhou and Yunnan	2006–08	Promote bilateral cooperation in clean technology, development of CDM projects	Financial support for four CDM centers, training
Development of Sino-Italian CDM Projects	Ningxia	2006	Development of three CDM projects	PDD development
China–Canada CDM Capacity Building	Hebei, Shanxi, Zhejiang, Shandong, and Hunan	2007	Exploration of potential CDM projects and development of PDDs	Financial support for CDM centers, training of local project owners and experts
China-Japan Shandong CDM Capacity Building Program	Shandong	2007–08	Development of CDM projects	Financial support for Shandong CDM center, training of local project owners and experts

Continued

Table 4.2 Continued

Donor	Location	Time	Objective	Activities
Sino-Danish Cooperation Project on CDM Capacity Building	Hunan, Guizhou, and Xinjiang	2007–08	Testing of the programmatic CDM approach	Pilot biomass projects
China-UNDP Millennium Development Goals (MDGs)	Twelve western provinces	Start in 2007	Development of CDM projects in western provinces which have a measurable contribution to the MDGs	Training for twelve provincial CDM centers, possible establishment of carbon-exchange platform between sellers and buyers

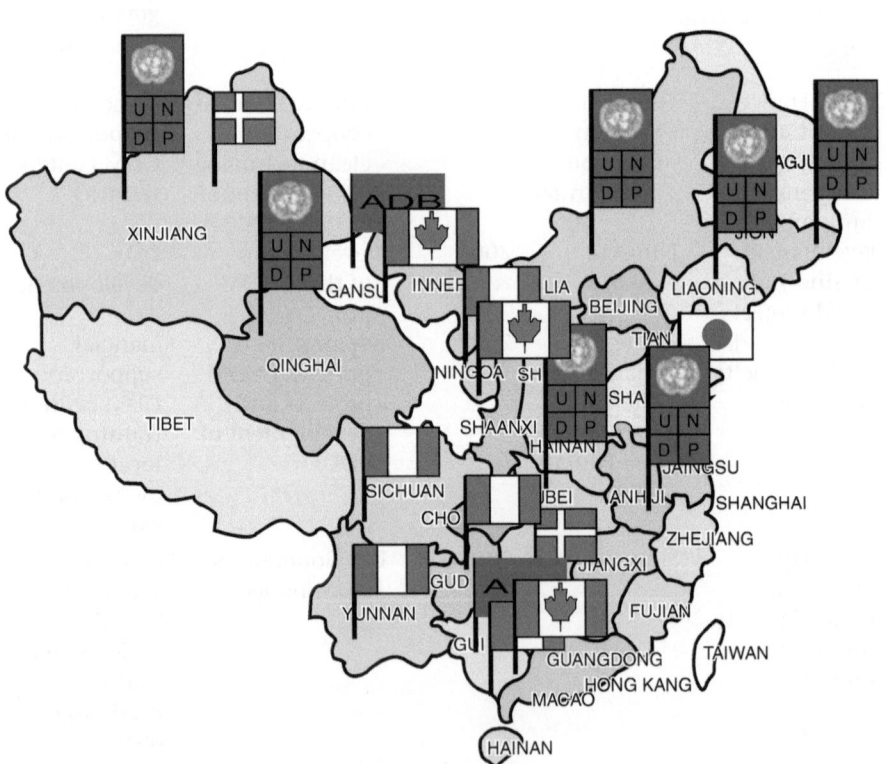

Map 4.1 Examples of Sino-foreign CDM capacity development projects at the provincial level

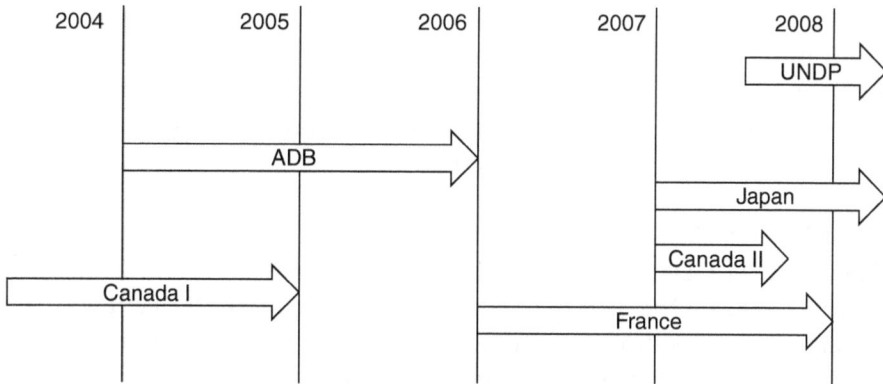

Figure 4.1 Timing of donor programs supporting CDM centers

Canadian companies did not enter into ERPAs, due to a domestic political turnaround that made their Kyoto obligations obsolete.

The CDM center in Yunnan received its seed money from a Sino-French cooperation program which supported the establishment of CDM centers in the four south-western provinces of Yunnan, Sichuan, Guangxi, and Guizhou, as these are in the part of China in which France has a historically strong presence and many Sino-French joint ventures today. In addition, these provinces all belong to the less-developed provinces that are part of China's Western Development Strategy. The program supports each of the four provinces with €175,000 for the establishment of their CDM centers. The program in Yunnan was launched with a inaugural workshop on February 9–10, 2006 in Yunnan's capital Kunming, and lasted until June 2008. It included training for the CDM centers' staff, and business networking events with French companies that were allegedly granted the first right to purchase CERs from CDM projects developed by the Yunnan CDM center (Interview 64).

Due to the broad similarities in program implementation, the impact of the donor programs on the CDM centers' effectiveness can be assumed to be similar, although their different timings might influence their effectiveness. There are also differences in the financial support CDM centers receive from foreign donors, but these seem not to correlate with the effectiveness of the CDM centers: the Ningxia CDM center was supported with €99,186, the Gansu CDM center and the Hunan CDM center each with approximately €50,000, all three by the Sino-Canadian program. The Yunnan CDM center received considerably more financial support, namely €175,000 from the Sino-French program. There was thus no great financial difference, although the

Ningxia CDM center also receives substantial extra financial support from Italian and British research programs. Although the Sino-French program provided three times as much support to the Yunnan CDM center, at the time of writing this financial advantage was not yet influencing the effectiveness of the center.

The establishment of provincial CDM centers

The Ningxia CDM center was the first of its kind when it opened in October 2003. One striking aspect of the Ningxia CDM center is the crucial role of its director, Zhang Jiesheng. Apparently, upon realizing the good CDM potential of Ningxia, he talked to MOST, which in turn approached the Canadian DFAIT for support. Because Canada at that time was eager to support the CDM process in China, the Sino-Canadian pilot project was launched to support the establishment of China's first CDM center. Two-and-a-half years later, the first CDM project developed by the Ningxia CDM center was registered. The Gansu CDM center was opened in November 2005, shortly after the Kyoto Protocol had come into force. Within a year, the center was able to register the first CDM project it had developed itself. The Hunan CDM center was also established in mid-2005, and the Yunnan CDM center was launched in the beginning of 2007.

Organizational strength

The four CDM centers show differences in their organizational strength that are determined largely by the type of ownership of a center, but also by its degree of political support and resource endowment. If the organizational strength of the CDM center according to these factors is high, one can expect this strength in turn to increase the effectiveness of the center's interventions.

One of the fundamental observations about the Chinese CDM centers is that they do not represent a single organizational model. Instead, they differ significantly in their political independence, their financing, and their ownership. Ownership models range from 100 percent government ownership, through centers that have both a public and a private entity (of which the government might be a shareholder), centers that are privately run entities serving only the government (*shiye danwei*), to centers that are private companies (again taking various forms – for example, joint stock companies or joint ventures).

The Ningxia CDM center is a typical example of a Chinese company with 'two heads but one body.' The staff and its director belong simultaneously to the public Ningxia CDM center and to the private Beijing Keji Consulting (see Figure 4.2). Although these are legally two different entities, the private Beijing Keji Consulting holds a 90 percent financial share of the Ningxia CDM center. At the same time, the Science and Technology Department has executive authority over the center. A representative of the center explains this complex situation:

> The Ningxia CDM center is a private entity and independent agency, a profit-making company with limited liability, but the Science and Technology Department is its instructing authority, although the center is not part of it. (Interview 42)

In addition to its attachment to the provincial government, the CDM center also has close ties with the purely private consultancy Shanghai Chuanji, which was co-founded by the center's director, Zhang Jiesheng. However, according to him, the private consultancy Shanghai Chuanji only has normal business relations with the CDM center and with Beijing Keji Consulting, and has separate staff, finances, and decision-making structures. The relation between the CDM center and Beijing Keji Consulting is much closer because they share the same staff and director.

The Gansu CDM center, on the other hand, represents an entity that is clearly part of the provincial government. This center is a *shiye danwei* attached to the Promotional Center for Scientific Development, and has

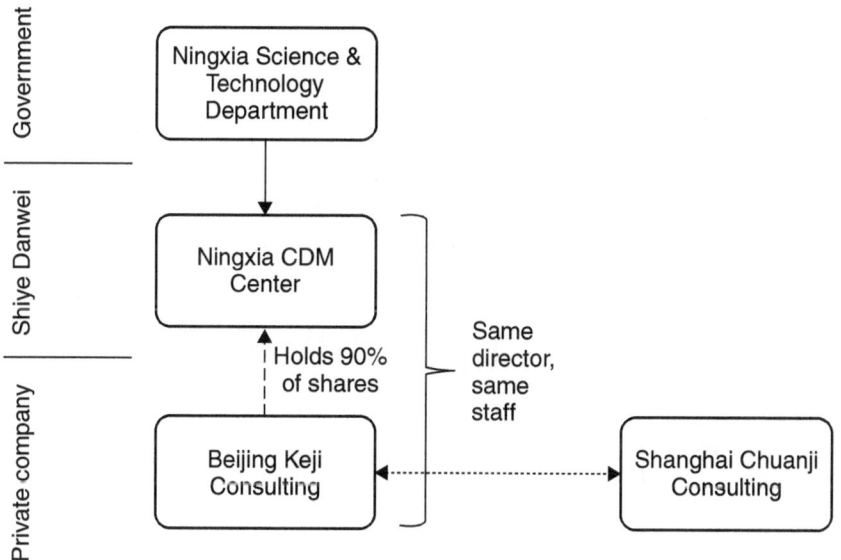

Figure 4.2   Structure of Ningxia CDM center

a bureaucratic rank equaling that of a county government (Interview 50). Institutionally, the Gansu Science and Technology Department is its political lead organization, and the Gansu Center for the Progress of Scientific Development is its supervising organization (see Figure 4.3).

During the discussion with the CDM center's director, Liu Jin, it became clear that the Gansu CDM center was much more dependent on the support of the provincial government than its counterparts in Hunan and Ningxia. The main reasons for this were its ownership structure as a collectively owned and financed center and the government tasks that it had to fulfill. This center was thus much more constrained in its decision-making, especially when it came to the question of the activities to which it devoted most resources and staff:

> In our setting the collective part is larger than the individually owned one, so that we cannot make decisions alone and have additional tasks beside the CDM. If we would concentrate only on the CDM, we could do much better... It is all a question of market and government. For Hunan and Ningxia, the market matters most and they do not face many political restrictions. For Gansu, we first have to think about everything from the government's perspective, and then we think about the market aspects. Making money is only a side aspect for us; instead our first task is to spread CDM knowledge among the companies of Gansu. For [Hunan and Ningxia] the market comes first. (Interview 49)

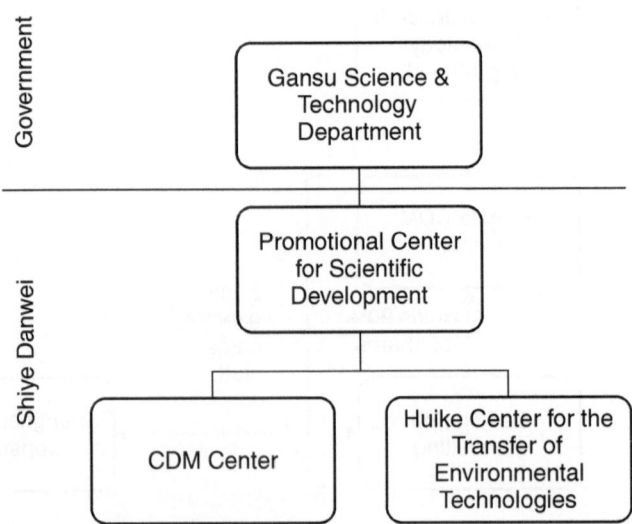

Figure 4.3   Structure of Gansu CDM center

The analysis of the four CDM centers alone reveals a plethora of possible organizational arrangements: the CDM centers in Gansu and Yunnan constitute parts of the provincial administration, while the CDM center in Ningxia comprises both a public entity and a private consultancy, and the Hunan CDM center is merely a private consultancy with public backing (see Table 4.3).

These diverse ownership and operation models exist because there is no central government directive on how provincial CDM centers should be configured. Instead, this decision is left to the provincial government, which can decide which entity to turn into an official CDM

Table 4.3 Comparison of CDM centers' institutional structure

Institutional structure	Example	Description
Part of provincial government only	Gansu, Yunnan	The CDM centers are a subunit of a *shiye danwei* that is part of the provincial government. The provincial government (the Science and Technology Department and/or the Development and Reform Commission) has instructive power on decision-making and provides financial and political support. CDM centers tend to have broad mandates with several non-CDM-related objectives.
'One entity with two heads' – one being part of government, the other being a private company	Ningxia	The CDM center is a subunit of a *shiye danwei* that is part of the provincial government. At the same time, the staff of the CDM center also work for a private consultancy which is headed by the CDM center's director. While the public unit may take over non-CDM-related tasks, the consultancy is focused on CDM project development.
Private company with government affiliation	Hunan	A private company is the official CDM center. It works for profit and concentrates on CDM project development, but must also respond to requests from the provincial government in order to ensure political support and access to financing – for example, from Sino-foreign donor programs.

center. When asked about the public and private dimensions of the provincial CDM centers, a representative of MOST explained:

> CDM centers are all different in each province. According to my knowledge, no fully private company exists; all have some kind of affiliation with the Science and Technology Department, the Development and Reform Commission, or both, depending on province. In the French[-supported] provinces it's the DRC, in Qinghai [UNDP-supported], it's the Environmental Protection Bureau. (Interview 1)

The ownership model of a CDM center directly influences its political standing. The ability to exert political influence is in turn a crucial factor determining the center's strength. The support the center receives from local political leaders reflects the organization's political standing and its embeddedness in the province's power relations and political networks. Its political standing depends also on individual political leaders and their support for the center, which can be said to be based partly on (personal) relations with the center's leaders, and partly on provincial concern with emission reductions. For the former, the Ningxia case study is a good example, as the center's director used to be the director of the Ningxia Science and Technology Department, and thereby one of the most senior political leaders of the autonomous region, until his launch of the Ningxia CDM center. With such strong political standing, it proved to be easy to secure government support for the nascent CDM center.

Another illustration of the importance of political backing for a CDM center's performance is provided by the Yunnan CDM center. Compared with the other three provinces, the attitude of the provincial government towards the Yunnan CDM center seems strikingly neutral. One reason might be that the Yunnan center is not the only entity within the Yunnan provincial government that has an interest in being the official CDM center. Instead, there is an interdepartmental competition to become the province's CDM focal point: there is the 'official' Yunnan CDM center, which is attached to the province's Development and Reform Commission, but there is also a second Yunnan CDM center, which is part of the local Environmental Protection Bureau; in addition, the Yunnan Forestry Department is also developing own CDM activities in its sector. A consequence of this competitive constellation seems to be a lack of coordination between various government departments, which in turn hampers the effectiveness of the Yunnan CDM center.

The type of ownership also influences a center's staff incentive structure. Governmental units such as the Gansu and Yunnan CDM centers have a broader mandate that also includes more non-CDM-related tasks than their private counterparts, which tend to concentrate on profit-making CDM project development. A broader mandate usually means that less time and resources can be devoted to CDM-related activities. In contrast, a more private structure also provides more monetary incentives for the centers' staff for good performance in CDM project development. For example, the director of a private CDM consultancy is able to reap profits directly, and thus has a high motivation to develop profit-making business strategies. These monetary incentives are often passed down to staff, who receive extra bonuses and company shares in return for good job performance. In contrast, CDM centers that are units of government, like the one in Yunnan, often have to work with government employees who already enjoy a lifetime government post, and are not allowed to receive extra payments based on job performance.

The human resource endowment of the centers – meaning the capacity and commitment of their staff – is also important for the centers' effectiveness, especially in relation to developing high-quality CDM projects. The four centers differ in terms of the number and capacity of their staff. These differences can be explained by the different salaries the centers can offer. The Yunnan center simply transferred staff from other governmental departments, requiring no special CDM knowledge and experience on their part, but also not offering them any additional financial incentives. The Gansu center relied partly on external expertise, using researchers from the nearby Lanzhou University, for example, as trainers in their CDM training and for PDD development. The Hunan center, in contrast, recruited new personnel who had to have CDM-related educational backgrounds and foreign-language skills, but who were also offered salaries higher than those for governmental positions. Nevertheless, it was the Hunan CDM center that was suffered particularly from high staff fluctuation. Reasons offered for this included the good job prospects in the Chinese carbon market, with other (international) private consultancies offering still better conditions, often including postings in Beijing. This point was also raised several times by interviewees in my research – representatives of CDM centers and private project developers alike. The problem of high staff turnover is experienced not only by the CDM centers, but also by the representatives of national and international private consultancies: 'So people trained at the provincial level or in public entities might leave for large private consultancies in Beijing' (Interview 18). Another reason given for the high staff turnover on the CDM market is the alleged preference of Chinese employees for government jobs over private-sector jobs. For example, since the Hunan CDM center is seen as a private company, people prefer to take secure government positions when they have the opportunity. Providing material and non-material incentives for staff is one fundamental means of developing and retaining high-quality advisory services at CDM centers. This finding is in line with recommendations in the literature on institution building relating to how to increase staff performance and loyalty by providing incentives and attractive career prospects (UN, 1982: 31).

With regard to organizational sustainability, it is clear that government affiliation can be positive for organizational sustainability because governmental tasks can become crucial for organizational survival in times of low market demand, but can also divert organizational focus from CDM activities. Interviews reveal that the future business development strategy of a CDM center is mainly dependent on its institutional formation and its mandate. The almost fully privately owned Hunan CDM center, for example, has very ambitious long-term goals: one representative has clearly voiced his interest in going beyond business-as-usual CDM project development to become active in CER trading, the market for voluntary emission reductions (VER), providing CDM capacity development in other provinces and abroad, and developing the center's own CDM methodologies. While Hunan and the Ningxia CDM centers are clearly inclined towards continuing their CDM business, or even expanding their services, the public Gansu CDM center has to deliver many non-CDM-related tasks, and thus has to divert its resources and efforts into research, policy advice, and other activities that are not even climate-related.

In addition to factors that have been identified prospectively based on innovation theory and prior practical experiences, two further important factors for explaining the centers' performance can be discerned: the type of leadership of a center and its perceived trustworthiness.

Type of leadership is one important internal factor determining the organizational strength of a CDM center. Because CDM centers are small entities and enjoy a relatively high degree of operational independence, the persons serving as the centers' directors can retain substantial influence on the overall performance of the organization. Their political standing, their entrepreneurial attitude, and their identification with the centers' mandates each make a difference. For example, as the director of the Ningxia CDM center was allegedly the person who came up with the initial idea of setting up provincial CDM centers, and has since received much praise and probably financial rewards, Zhang Jisheng is a very determined leader of the organization and its affiliated private consultancy. The director of the Gansu CDM center, Liu Jin, represents the typical Chinese cadre: he served in the provincial administration long before becoming the director of the center, another step up on the political career ladder. The director of the Hunan CDM center, Zhang Hanwen, is the only one of the four directors interviewed who had not worked in the government before. Instead, he had occupied leading business positions and was able to transfer his business leadership skills to the work of the Hunan CDM center, which resembles a privately run company. The question of how much individual leadership determines organizational performance is of course a contested issue, as people consistently overestimate the effect of leaders' personalities on organizational performance (Pfeffer and Salancik, 1978: 7). While leadership certainly depends partly on personal character, there are structural causes that can motivate good performance. These include, for example, financial awards, performance-based contracts, and an organizational culture emphasizing promising ideas and practices and offsetting the risks of innovation (Campbell and Fuhr, 2004: 4, Moore et al., 1995: 50). Part of the leadership phenomenon found in the CDM centers can also be explained by the Chinese cadre-system and its incentive structure. It is, for example, very common to have top-level officials heading 'private' enterprises during periods of economic transition. While close connections to the government apparatus proved to be helpful for the CDM centers, this additional role of political cadres as business executives also opens up potential opportunities for rent-seeking: 'These bureaucrats now can make use of fractures in the planning system and imperfections in the market to capitalize on whatever assets and connections they command, earning extra commissions as they do.' (Solinger 1992: 135)

Trustworthiness has proved an important factor in persuading companies to turn towards CDM centers in situations of uncertainty. The case studies revealed that the trustworthiness of a CDM center was closely related to the degree of its governmental affiliation. Centers that were closely affiliated with the provincial government, or even part of it, were more trusted by potential project owners than centers that operated independently. This statement holds, however, only for the early phases of the provincial CDM markets, and could even be reversed for the advanced phases, in which private entities are trusted more due to their professionalism (as in Yunnan). In their early stages, CDM centers are perceived as especially trustworthy if they are attached to the provincial government (this point was made in Ningxia, Gansu and Hunan). Government backing makes them trustworthy, because 'they will not just disappear tomorrow' and 'they do not just want to make money.' Another explanation offered by an interviewee was that Chinese companies are overwhelmed by the international market, do not know how to deal with international companies, and thus look to the government for guidance (Interview 63). For example, project owners in Gansu decided to cooperate with the CDM center because of its trustworthy position as a subordinate part of the government's Science and Technology Department, although they had previously been approached by other private project developers.

This situation was different in Yunnan, where the majority of project owners rely on private consultancies for CDM-related information. Although the Yunnan CDM center forms part of the provincial government, it was not perceived to be as trustworthy as its private competitors. Interviewees stated that, although they had heard of the Yunnan CDM center's existence, they either know nothing about its activities, or, if they had, judged them to be ineffective. Due to this perceived ineffectiveness of the Yunnan CDM center, the project owners who were interviewed preferred to cooperate with private project developers. Neither, apparently, could the Yunnan CDM center profit much from the trust that other provincial CDM centers had gained from their government affiliation. One project owner explained why:

> They [the private consultancy] explained the CDM very explicitly; nobody else explained it that clearly. I would rather trust a private company than a SOE. There are private companies in Yunnan, but they also did not explain that very clearly. If they have a successful example, I would also trust them, because if they have more experience in doing CDM, the projects will be more efficient and will save time. (Interview 71)

Project owners placed a high value on the ability of private consultancies to explain the CDM in clear and understandable terms. This kind of expertise seems to be much more important in their decisions on who to cooperate with for CDM project development than the mere trustworthiness of a government institution with little real CDM expertise. A reason given for their preference of private consultancies is that information offered by private (mainly Beijing-based) consultancies was qualitatively better than the information provided by the CDM center, and was provided at an earlier stage. Departments of the provincial government were seen as operating without coordination, information, or efficiency.

My findings confirm the factors identified in advance that were relevant in determining a center's organizational strength, and point to further relevant factors – the type of leadership and the perceived trustworthiness of a CDM center. These findings revealed, however, that those factors are not equally important in determining the centers' organizational strength, but that the ownership model of a center influences the value of the other factors, such as incentive structure, resources, organizational sustainability, leadership type, and trustworthiness. The ownership model (private, public, or a mix thereof) has thus proved decisive for the CDM centers' organizational strength, because it has a direct influence on the other indicators of the organizational strength of a CDM center. With regard to the influence of these factors over time, a government affiliation is seen as very positive for the early phases of a CDM market, because it engenders trust in potential project owners in times of insufficient information and capacity. In later CDM market phases, however, government affiliation tends to affect a center's organizational strength negatively, because it places limits on incentive structures, resources, and business development strategies.

CDM centers' effectiveness in CDM promotion

All four CDM centers have mandates to spread information about the CDM to other market actors in their province. All four centers have therefore employed strategies for increasing such awareness, some of which were common to them all, though varying in scope, while some were only deployed by one or two CDM centers. All four CDM centers have produced brochures and introductory booklets about the CDM. These materials are then either mailed to potential project owners, handed out during conferences and workshops, or distributed through the information channels of provincial governments.

In the theoretical expectations developed in Chapter 4 about the impact of CDM centers on provincial market development, it was anticipated that the use of mass media for information dissemination would influence CDM awareness within a province. Indeed, all four CDM centers under examination used the internet to disseminate CDM information. All centers launched their own websites providing information about the mission and services of the CDM center, provincial market development, national market development, CDM regulations, and international CDM market news. The timing of the website launch of each center differed: the centers in Ningxia, Gansu, and Hunan set up their sites in 2006, but the Yunnan center only launched its site at the beginning of 2008. Not surprisingly, the reception of the websites also differed by province. While several interviewed project owners in Ningxia, Gansu, and Hunan were aware of the center's website and used it for information on the CDM, and two project owners from Ningxia and Hunan even stated that it had 'great impact' on their work, no interviewee knew about plans of the Yunnan CDM center to set up a webpage. Although the centers' websites shared many similarities in their basic structure, the website of the Hunan CDM center was the most user-friendly, providing China's first-time online chat tool for project owners to inquire about the CDM, and offering a free download of the guidelines for an initial CDM eligibility test. The CDM centers also used other forms of media to spread information – for example, they invited journalists to write CDM-related articles in local newspapers or to produce relevant documentaries for the provincial radio and television channels.

The expectations about factors determining an agency's effectiveness in disseminating information relate to the clear definition of a target group, the type of information provided, and the type of communication channels used. While the latter factor proved important for effective awareness-raising, the target groups and the type of information provided by the four centers were similar, and thus had no particular influence. Information disseminated in public media targeted the general public, while CDM conferences and official governmental information leaflets targeted government officials and potential project owners. One finding that contradicted the expectations was that no CDM center had targeted other project developers in their information dissemination activities, because it perceived them as competitors. Moreover, the type of information provided was homogenous. For example, CDM centers also publicized CDM handbooks to complement training. These handbooks were similar in their content, and usually included an introduction to climate change, the Kyoto Protocol and the CDM, an overview of current Chinese CDM institutions, regulations, and registration procedures, and an outline of the CDM potential of sectors of the province concerned. In addition, information was provided on Chinese CDM regulations and project approval criteria. The CDM was usually framed as a means of financing emission reductions, thereby falling in line with the central government's policy on 'energy savings and emission reductions.'

An unexpected strategy of the CDM centers in promoting the CDM was the very effective use of internal channels of the provincial government to reach out to potential project owners, informing them about the CDM and lobbying them to become active in developing their own CDM projects. For example, the Ningxia CDM center was able to use the official communication channels of the Science and Technology Department to reach out to potential customers in various sectors and on all administrative levels. By being able to frame the center as a government entity (although it was a partly private company) and by using the official communication channels, project sourcing was made easier. The center's deputy director explained how he was able to make use of the government's communication channels. When he worked in the province's department for industry technologies, he was in frequent contact with companies with high GHG emissions. After switching jobs to the CDM center, he was still able to use his connections from the previous job (Interview 42).

An attachment to the provincial government allows the CDM centers to have either the Science and Technology Department or the Development and Reform Commission issue official invitations to their CDM conferences and workshops. This works well because governmental officials at the district, city, and county levels can directly contact project owners who they know have good potential for GHG emission reductions. These project owners are then invited by government officials to the training sessions and workshops of the CDM centers (Interview 64). The use of government communication channels enables the centers to expand their activities to all levels of the province. This vertical outreach to provincial sublevels would be very hard for private companies to achieve, as they would need offices on the district, city, and county levels to have the same outreach capacity (Interview 57).

All CDM centers also used interpersonal communication strategies to communicate information on the CDM during provincial-level CDM conferences. These normally included high-level government officials like the vice-governor or the directors of the province's Science and Technology Department and the Development and Reform Commission. These high-level government officials were asked to express their support for the CDM and the CDM center by, for example, delivering remarks welcoming those attending. On average, such first-time CDM conferences were attended by around 100 people. High-level provincial politicians and/or well-known scientists were often invited to attend these conferences and to express their support for the CDM (Interview 3). In Gansu, for example, Yang Zhiming, vice-governor of the province, displayed his support for the CDM by speaking on 'The Circle Economy as a New Path for the Western Development Strategy'; and Zhang Tianli, director of the Science and Technology Department, spoke on 'Using the Momentum to Realize CDM Projects in Our Province.' According to the CDM center's representative, this conference had a deep impact on CDM awareness in Gansu, because Yang, vice-governor of Gansu province, displayed his support for the CDM by linking it to the concept of a 'circular economy.' In addition, the vice-governor also introduced a five-step model for the use of the CDM to strengthen international cooperation, which is now used by other Chinese CDM centers. Because CDM conferences targeted the province's decision-makers, one can verify that CDM centers adhere to the principle of initially targeting opinion leaders. It is thus opinion leaders who back the CDM, and thereby create an atmosphere of trust among potential adopters. What were the effects of the information dissemination strategies employed?

The CDM centers' information-dissemination strategies increased CDM awareness in their respective provinces to different extents, depending on their timing and on the group they were targeting. Through peer assessment, an increase in CDM awareness was attested by the interviewees for the three groups, comprising a) potential project owners, b) government officials, and c) financial institutions. The quantitative results of the interviews reveal that an increase in CDM awareness occurred in all four provinces, but that it varied in time and depth (compare Figures 4.4, 4.5, and 4.6, which give, on a scale moving from

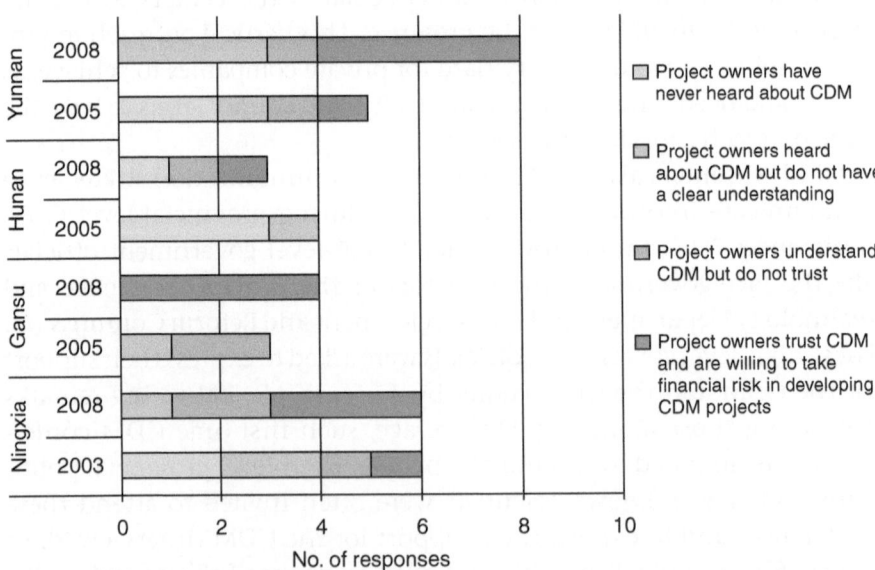

Figure 4.4 Comparison of change in CDM awareness among potential project owners

light grey to dark grey, the differences in CDM awareness. The size of the bar indicates only the number of responses received).

CDM awareness among potential project owners increased in all four provinces from 2005 to 2008. Many potential project owners had already heard about the CDM in 2005 (in Ningxia, as early as 2003), but project owners did not yet believe in the CDM. Hesitation and distrust were widespread, but awareness of the CDM and a willingness to become active were steadily increasing. In 2008, only one interviewee stated that project owners in Hunan had not yet heard about the CDM at all. In the other three provinces, at least half of all project owners knew about the CDM, understood and trusted it, and were willing to take financial risk in developing CDM projects.

Reasons for the increase in CDM awareness among project owners include activities of the CDM center as well as other information sources:

- *Media and the internet.* All project owners interviewed stated that – in addition to the CDM center's information – they used the internet to search for background information on the CDM. This would have included the NDRC website and the UNFCCC website, as well as other information sources. In addition, Chinese newspaper and television channels also broadcast reports on climate change and the CDM. Information on the CDM was available in a Ningxia newspaper as early as 2002/03 (Interview 45). Many project owners heard about the CDM for the first time in 2005 from the television and local newspapers. Internet resources such as the UNFCCC website, the NDRC website, the CDM centers' websites, and Google were also important sources of information. English websites were seldom used due to language barriers.
- *Conferences and workshops organized by the CDM center.* Some project owners also attended the workshops for potential project owners organized by the CDM centers, which introduced the CDM and the project eligibility criteria. The purpose of the conferences organized by the CDM centers was to introduce the main functions of the CDM, and to make the eligibility criteria for CDM projects known in order to build trust and understanding among companies so that they would start CDM projects.
- *CDM publications from the CDM center.* Project owners stated that the CDM information material received from the centers when attending training sessions provided some useful information.
- *Provincial government propaganda.* Some companies heard about the CDM for the first time due to an official notification sent out by the provincial government. The Gansu provincial government paid

special attention to the CDM in its propaganda – for example, in its party newspaper (*Dang Bao*), which is read by many private companies as well as government officials. According to one project owner, this government backing of the CDM contributed decisively to increasing the trust of project owners in a mechanism of which they were initially skeptical (Interview 54).

- *Conferences organized by NDRC.* Several project owners stated that – in addition to the CDM conferences organized by the provincial CDM centers – they also attended CDM conferences in Beijing organized by the NDRC. Although these conferences were perceived as very trustworthy because consolidated information was provided by the NDRC, some project owners complained about the exorbitant fees.
- *Industry associations.* There was also a peer-to-peer information exchange initiated by sectoral industry associations – for example, by the Cement Industry Association (*Shuini Xiehui*) (Interview 45). An informal information exchange between peers also took place among hydropower project owners, so that almost all hydropower companies would then use the CDM if their projects were eligible.

CDM awareness among government officials also increased in all four provinces. Representatives of the four provincial governments, however, showed quite different levels of CDM awareness at the beginning of the CDM market in China. For example, Ningxia's government officials already had exceptionally high CDM awareness in 2003, which also rose exceptionally high in 2007/08 in comparison to that of Yunnan and Hunan. Only government officials in Gansu reached a comparable level

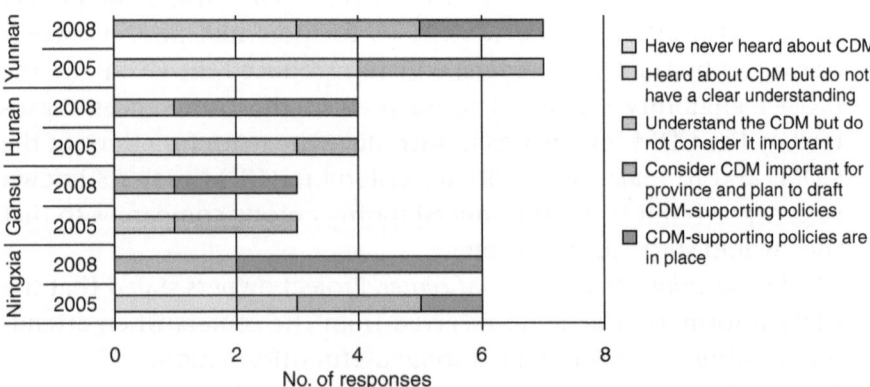

Figure 4.5 Comparison of change in CDM awareness among government officials

of awareness in 2007/08. CDM awareness among Yunnan government officials increased, between 2005 and 2007, from almost no knowledge to understanding the idea, but uncertainty about details. According to the interviewees, it was also important to distinguish between the different entities of the provincial government. Usually the CDM knowledge of officials from the Science and Technology Department and from the Development and Reform Commission was better than that of officials of other departments, because the former had participated in several CDM workshops and had been actively involved in policy-making processes relating to the CDM. Reasons given for an increase in the awareness of government officials were:

- *Central government directives.* An increase in CDM awareness could be due to the central government's policy on 'energy saving and emission reductions.' Since the launch of this policy in 2005, government officials at all levels were supposed to be aware of the issue of climate change, and to develop an interest in mitigation measures like the CDM.
- *Information from the NDRC and MOST.* The NDRC and MOST gave early support to the CDM, and spread relevant information to the provincial Development and Reform Commissions and the Departments of Science and Technology, which were asked to take up the issue.
- *CDM conferences organized by the CDM center and their donors.* Government officials from the four provinces were also invited to participate in several CDM conferences, which were organized by the CDM centers and their respective donors.
- *Media and the internet.* The mass media (television, newspapers, and the internet) were also initial providers of information on the CDM and the international climate change discussion to government representatives.

Financial institutions in China were completely unaware of the CDM and its profitability at the beginning of the CDM market. At least one interviewee in each province besides Gansu stated that financial institutions had not heard about the CDM in 2005. This assessment changed for 2007/08, when interviewees stated that financial institutions had heard about the CDM but either did not trust it or would not consider it for their loan assessment. The CDM awareness situation was again exceptional in Ningxia and Gansu, where several interviewees stated that banks would consider the CDM revenues for their loan assessment if the project owner was able to provide an ERPA contract with a foreign company. One interviewee even stated that investment banks in

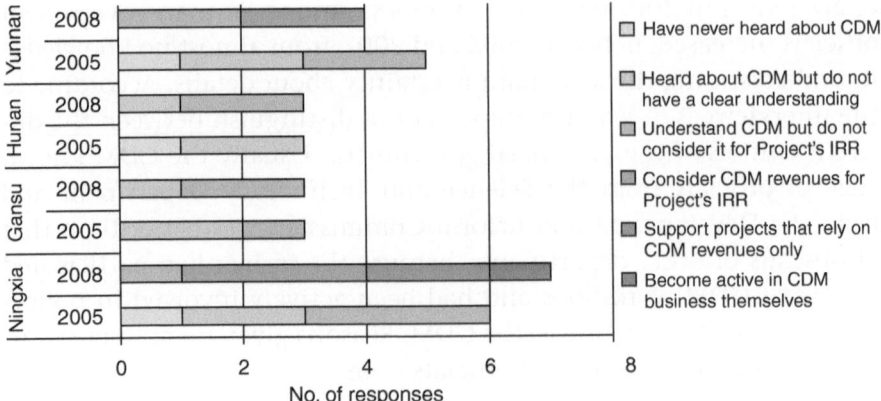

Figure 4.6 Comparison of change in CDM awareness among financial institutions

Ningxia would be active in the CDM business themselves. The following reasons were given for the change in awareness among financial institutions:

- *Successful CDM projects.* A change in attitude occurred as a result of the banks' realization that CDM projects made good revenues and enabled companies to pay back their loans earlier (Interview 60).
- *Information and lobbying by CDM centers.* According to one project owner, the change in attitude was due to the 'good work of the center' in relation to information dissemination and lobbying (Interview 60).
- *Explanations by project owners.* Often, companies wanting to undertake CDM projects first had to learn about the CDM themselves, and then explain its function to the local banks they approached for loans (Interview 61).
- *Lack of trust.* In the early market, financial institutions did not trust the CDM. Interview partners, however, were convinced that this attitude would change as soon as the first successful CDM project showed profitability and received CERs issued through the international transaction log (Interviews 58 and 59).
- *Focus on profitability.* Although the China Banking Regulatory Commission (*Yin jian hui*) issued a decree for all banks to give priority to environmental projects, banks in reality judge projects by their internal rate of return (IRR) (Interview 45).
- *Inability to become active on the carbon market.* Chinese financial institutions cannot become active as traders on the carbon market, which has severely limited their interest in the CDM (Interview 46).

But interview statements revealed some frustration with the early and persisting unwillingness of banks to take the CDM into their loan considerations. Project owners still considered it very difficult to obtain loans for their CDM projects (Interviews 81 and 82). Interestingly, it is the project owners themselves who need to explain the CDM and its revenue flow to the banks when they request a loan (Interviews 75 and 81). According to two project owners this is a very difficult job, because project owners have no means of proving the reliability of CER revenues, since only a few CDM projects have so far been issued CERs, though many are already registered. One possible explanation for the reluctance of Chinese local banks to consider CDM revenues in their loan assessments is the state control of the Chinese banking system:

The problem with financial institutions in China is that they do not do investments. Until now the Bank of China was not active in the CDM. They are not comparable to the foreign banks which pick up the CDM business very quickly. There is not one [Chinese] bank that has engaged in investing in or developing CDM projects so far. There are now a few banks which have regulations regarding the CDM, because the CDM revenues can reduce the project risk in the long term, so the CDM is considered positively for loan assessments. (Interview 56)

In addition, local bank branches have to abide by the central regulations. In order to achieve some change in attitude, CDM center staff would therefore need to go to Beijing to reach the central bank decision makers, as the director of the CDM center explained (Interview 56). The unsupportive attitude in the banking system might be one reason why representatives of financial institutions have not yet participated in CDM training (Interview 56).

These results in levels of CDM awareness among different actor groups show that, despite steady national and international CDM market development, which should have had similar effects on all Chinese provinces, disparities in CDM awareness were present at the provincial level, and they seem to persist over time. Nevertheless, time is a crucial factor in explaining the centers' effectiveness in CDM promotion. The case studies show, for example, that the Ningxia and Gansu CDM centers' information-dissemination strategies were important for raising CDM awareness in the early phase, while the Hunan and Yunnan centers' information-dissemination activities were launched at a time when provincial market actors had already heard of the CDM from other international, national, and even provincial sources. Although the Hunan CDM center has a good website, most market actors 'had already chosen their preferred information sources before the Hunan website became known. Interestingly, none of the CDM centers was able to establish a website that was ranked by interviewees as equal to or better than those of the Chinese NDRC or the UNFCCC. Interviewees preferred these two websites to all others because they provided consolidated and reliable information.

The information-dissemination practices of CDM centers showed that the type of communication channels used proved decisive, while almost no differences in the centers' strategies could be detected in relation to their target group or information content. The case studies also revealed the effective use of websites and government communication channels in reaching out to potential project owners. The impact of these communication channels varied over time, however: they were important in the early market phases, when no other information source was available and when opinion leaders from the provincial government could establish trust in the CDM through their support.

CDM centers' effectiveness in CDM capacity development

CDM capacity development is one of the core mandates of the provincial CDM centers, especially because most of them started their operation at a time when knowhow and capacity relating to the CDM among potential market players was very low. The creation of learning opportunities should be the main strategy of the CDM center in increasing CDM capacity. Project owners should develop the ability to judge their project's CDM eligibility, and should become willing to enter into cooperation agreements with a CDM center or any other project developer. Government officials should provide support to the CDM by developing relevant policy measures. Financial institutions can be expected to attain enough CDM knowledge so that they are willing to accept the CDM as collateral.

CDM centers have indeed enabled a learning process among selected market actors by organizing CDM training sessions, but these differed in timing, frequency, and scope (see Table 4.4). Some only held the training sessions that were compulsory in the framework of their Sino-foreign projects, while others held self-financed training sessions in order to make contact with potential project owners and to source projects. It was common to all CDM centers that they held an inaugural conference/kick-off workshop (i.e. Category 1 of the training types

Table 4.4 Comparison of types of training conducted by CDM centers

	Ningxia	Gansu	Hunan	Yunnan
Number and type of training	Category a = 2 Category b = 5 Category c = 0	Category a = 1 Category b = 5 Category c = 1	Category a = 4 Category b = 6 Category c = 2	Category a = 1 Category b = 2 Category c = 0
Timing	2004–05	2005–07	2005–06	2007–08
Place	Provincial capital	Provincial capital and district	Provincial capital and district	Provincial capital and district
Target group	Provincial government, project owners, journalists	Provincial government, project owners, banks	Provincial government, project owners	Provincial government, project owners, researchers
Usage by interviewees	All but one	Only a few	Only a few	None

Category a = inaugural conference/kick-off workshop
Category b = CDM training
Category c = contract signing conference

outlined in Table 4.4), to which the donor agencies supporting the establishment of the CDM centers (CIDA for Ningxia and Hunan, ADB for Gansu, and AFD for Yunnan) were invited. These inaugural conferences followed a typical scheme that included speeches about the Sino-foreign CDM cooperation program, expert presentations on the technicalities of the CDM, and promotional presentations by government or private CER buyers. This type of training aimed at raising the interest of potential project owners and their willingness to cooperate in CDM project development. The project owners' confidence in the benefits of the CDM was increased by the presence of high-level government officials and international and Chinese experts from prestigious universities. The second type of CDM conference was a typical training session (Category 2 in Table 4.4) in which CDM experts made presentations on CDM procedures, conducted practical exercises for CDM project development, and sometimes even supported potential project developers in fleshing out initial descriptions of their prospective CDM projects. This kind of training was conducted either by the centers' staff or by invited (mainly Chinese) CDM experts. These training sessions targeted representatives of different provincial government departments and representatives of companies that had potential CDM projects in their jurisdiction or company. This category of training had the effect of transferring technical CDM knowhow to the participants. A third type of CDM event (Category 3 in Table 4.4) hosted by CDM centers consisted of 'contract signing conferences,' at which buyers and sellers signed the ERPA. For these conferences, high-level government officials and foreign representatives of donor agencies and companies were invited to express their support for the CDM. This category of training thus showcased the feasibility and profitability of CDM projects to project owners who were not yet convinced. These conferences served the dual purpose of closing existing deals and promoting new ones.

The timings of training sessions conducted show that the first CDM sessions always took place in the province's capital, whereas training was later also conducted at the district level. Interviews with project owners in the provinces revealed that all but one project owner in Ningxia had participated in the center's training and a few project owners had participated in the training in Gansu and Hunan, while none of the interviewed project owners had participated in training offered by the Yunnan CDM center. The Ningxia and Hunan CDM centers offered training only as long as they received assistance from the Sino-foreign capacity development programs. One project owner thus voiced a skeptical opinion about the CDM centers' contribution to capacity development. For him, the CDM centers generally contributed to CDM capacity development only in the early phases of the Chinese CDM market, but later on neglected this task and focused solely on profit-making project-development services:

> CDM Centers in Ningxia, Gansu, and Hubei were active in 2005 and 2006 for CDM capacity development and organized a few training [programs]. In addition, the NDRC has a CDM training center and also contributes to local CDM market development. After 2007, many people discovered the CDM, and there were many training [programs] held, but the CDM centers did not do training anymore because their eyes were cast on profit. They used to get money from the central government for their training; now they do not get money anymore, so they do no more training. Now they just look for new projects, develop projects, and make money. (Interview 47)

After the closure of Sino-foreign support programs, CDM centers either stopped providing training or shifted target groups towards the district level or towards sectors that were not yet popular CDM areas. Incentives to provide CDM training even after external assistance had been phased out were to spread CDM knowledge in order to get access to potential CDM projects and to find new customers for PDD development (Interview 3).

Training offered by the CDM centers was of mixed quality. It was usually perceived to be of higher quality if it was conducted by international or Beijing-based experts (mostly university researchers). The second-best solution was to have provincial researchers as trainers. Of doubtful quality was training conducted by the CDM centers' staff themselves, who might only have learned about the CDM shortly beforehand. Interestingly, these deficiencies mirror those that have been observed in training offered by public administration schools in countries in transition in Eastern Europe during the 1990s:

> A general problem with all the above-mentioned training institutions is that very often they are little more than managers of training programs. They do not actually have a core body of permanent trainers in their staff. Furthermore, they lack the capacities to carry out reliable analyses of training needs, often merely react to government and are dependent on foreign assistance for the provision of training programs in key areas ... (Verheijen, 2000: 31)

Notwithstanding a priori assumptions, CDM training organized by CDM centers did not target other project developers. While representatives of CDM centers always ensured that other project developers in their provinces were made welcome, they also stated that these would eventually become competitors. One fundamental conclusion is therefore that the CDM centers have had no impact on CDM capacity development for one important segment of the market: other project developers. Representatives of financial institutions were also strikingly absent from this CDM training. Representatives of CDM centers explained that they had been invited, but hardly ever attended trainings because their banks did not yet pay attention to the CDM. But there was some change in the provincial financial institutions' attitude towards the CDM, as some banks' headquarters started to issue guidelines on them for their local branches. In general, the lack of interest in the CDM on the part of Chinese banks can be explained by the low profitability of CDM projects in comparison to other types of local project, such as construction projects; the time-lag between current CDM project implementation and successful revenue flow once CERs have been issued; and the general tendency of Chinese banks to take up new business opportunities only after their central head office has given an instruction to that effect.

The capacity development interventions of the CDM center were fairly successful in raising the capacities of project owners and government officials, but unsuccessful in relation to other project developers. My empirical research shows that several of the interviewed companies participated in CDM centers' training and workshops in the early market phases. Learning about CDM procedures and project eligibility helped them in their decisions on whether to become active in CDM project development. Most project owners interviewed stated that they were interested in learning basic information on how the CDM functions and whether their project is eligible, but would leave the project development itself to experts. This is because the average project owner in China tends to be a small, local company without the necessary resources and time to acquire in-depth CDM capacity. Only large companies like the power utility group Datang can write PDDs themselves. Of the five power utility companies that exist in China, one (Guodian) has its own CDM Office and one (Huadian) is setting up an office, while the others cooperate with external project developers or universities (for example, the Huaneng Group cooperates with CDM experts from Tsinghua University). The same goes for other large industrial conglomerates – for example, the two large steel corporations Hubei Gugang and Shougang are cooperating with private consultancies (Interview 70). However, according to the representative of such a private consultancy, most of the big companies do not know much about CDM projects or procedures, and are often lost when incompetent or dubious project developers are unable to deliver good-quality services (Interview 70).

Interviews with the project owners of the four provinces reveal a correlation between the impact of training sessions and their timing. The earlier CDM centers managed to offer training in their province, the greater was their use to participants. If training was organized late in time, its impact on raising participants' CDM capacities was lower. The Yunnan CDM center, with its late market entry, faced just such a dilemma: due to the late launch of the CDM center and its training program, and due to the small number of CDM training sessions offered, the impact of that training on the CDM capacity of Yunnan market actors was considered by the interviewees to be low. None of the project owners interviewed in Yunnan participated in the training, but instead gained their CDM knowledge from other sources. Most project owners in Yunnan acquired their CDM knowledge from international and Beijing-based consultancies and from learning on the job. Even a representative of the Yunnan CDM center acknowledged that their CDM knowledge was only sufficient for developing small hydropower projects, while they themselves needed further training in order to be able to develop other project types (Interview 64). Interview partners in Yunnan evaluated the general CDM capacity of all actor groups – project owners, the CDM center, and the provincial government – to be very low in the peer-assessment exercise. This had led to a situation in which these groups needed to rely on international and Beijing-based consultancies to develop their CDM projects, even if 'these charge large amounts of money' for their services (Interview 73, Interview 70).

The companies that participated in training offered by the four CDM centers often subsequently became partners in CDM project development. This link between offering learning opportunities and winning new customers for PDD development was the major reason – besides the support offered by Sino-foreign capacity-development programs – for CDM centers to conduct such training. But there were also project owners who explicitly decided against participating in CDM training offered by CDM centers, because they doubted its quality and preferred to receive one-on-one training from private consultancies. Moreover, the NDRC itself turned out to be a direct competitor of the CDM centers, as it also offered CDM training in the capital and in the provinces, albeit at a high price. This competition between the NDRC and CDM centers as the provincial subunits of MOST reveals once more the rivalry between these two ministries in the area of CDM competencies.

The CDM centers and the provincial governments had a mutual relationship in informing each other about potential CDM projects. The normal procedure would be that someone within either the provincial Development and Reform Commission or the Science and Technology Department would be asked by the central government to select an institution to be the province's CDM center (the exception is Ningxia, where one government official apparently came up with the idea himself). In this early phase in the development of the market, information flows from the provincial government to its CDM center. Once the CDM center's staff have acquired their own CDM expertise through the Sino-foreign assistance programs and by other means, the center is able to start to organize CDM conferences and training sessions, to which government officials of various provincial departments would be invited. While the Ningxia and Gansu centers' training programs explicitly targeted government officials, the Hunan and Yunnan centers targeted companies, though also inviting government officials. Indeed there is some correlation for Ningxia and Gansu between a center's targeting strategy and the provincial government's positive attitude towards the CDM. But although the Hunan CDM center did not particularly target the provincial government, it was nevertheless favorable to the CDM. The Ningxia, Gansu, and Hunan governments all included the CDM in their long-term planning documents, and all three issued special government documents giving explicit support to the CDM. Interviewees also mentioned that the CDM and climate change were both issues high up on the political agenda. In 2007/08, the Yunnan government was also showing a positive attitude towards the CDM, but it had not yet included it in its long-term plans, though it was in the process of drafting its provincial climate change plans. The latter activity indicates that the CDM- and climate-related policies of the provincial governments are in some instances more influenced by instructions issued by central government than by capacity development and policy advice from CDM centers. Several Chinese provinces were requested to draft provincial-level climate change plans. This strong responsiveness to the requests of the central government regarding climate change issues in general, and the 'energy saving and emission reduction' policy in particular, was observed as a general trend for all provinces (Qi et al., 2008).

All four CDM centers thus offered CDM capacity development training. CDM centers offered CDM learning opportunities to their superiors (government officials) and to their potential customers (project owners), but refrained from addressing other project developers, who were perceived as competitors. In addition to the expectations outlined above, it has proved important to differentiate between the type, timing, and location of training sessions. CDM training, workshops, and conferences have helped to overcome the diffusion barrier represented by asymmetric information and high levels of uncertainty. By providing information about eligibility criteria for CDM projects and about the costs involved in the registration process, these capacity development measures should have enabled potential project owners to conduct individual cost-benefit calculations for whether adopting the CDM for their projects was economically worthwhile. But the learning opportunities offered by the CDM centers should also have stimulated the group dynamics of innovation adoption. The two targeted groups – project owners and government officials – had many opportunities to interact during training sessions and workshops, and to exchange their experiences of and opinions on the CDM directly. This process of learning by interacting should improve the learning curve among potential adopters and increase their willingness to make use of the innovation of the CDM. In contrast, opportunities for learning-by-doing were rarely offered by the CDM centers. Although successful pilot CDM projects proved to be important – because only after the successful registration of CDM pilot projects and the successful issuing of CERs have project owners developed trust in the mechanism – such projects were not purposefully set up by the centers. CDM centers offered opportunities to learn through practice only in the form of one-to-one CDM tutoring, in which they approached companies for project development.

CDM centers' effectiveness in project development

Although CDM project development is only one of their three main mandates, it is increasingly becoming the main focus of China's CDM centers. Besides the ability to earn profits from PDD development, developing and implementing CDM projects, if pursued successfully, can benefit CDM diffusion. Theoretical and practical experience from other fields, however, reveals two possible roles for market-facilitating centers such as the CDM centers. They can either develop projects directly, or facilitate development by other actors. The case studies revealed that all four CDM centers were active in providing direct CDM project development services; it eventually has become their core activity. Indirect CDM project development coaching did not take place, although some centers became active in lobbying their provincial government to support the CDM, thus easing project development.

The four CDM centers offered direct CDM project-development services, which were comparable to the services offered by private consultancies. These included project sourcing, writing of PINs, contract negotiations with the project owner, writing of PDDs, contract negotiations with buyers, match-making with DOEs for project validation and verification, and assisting project owners in winning DNA approval and EB registration. Service fees charged by the CDM centers were in the same range as fees charged by private consultancies or university researchers (Interview 27). The four CDM centers differed, however, in the scope and quality of the services they offered. Some centers were willing to undertake only a narrow range of CDM projects (for example, the Yunnan CDM center only develops hydropower projects); some centers outsourced more complicated project types to external developers; and some offered extra services (for example, the Hunan CDM center develops software for monitoring project emissions).

Staff capacity is one crucial factor determining the quality of the services offered by a center. Interviewees' assessments of the increase in CDM capacity of the centers' staff are shown in Figure 4.7. All four centers showed a lack of CDM project-development capacity at the beginning of

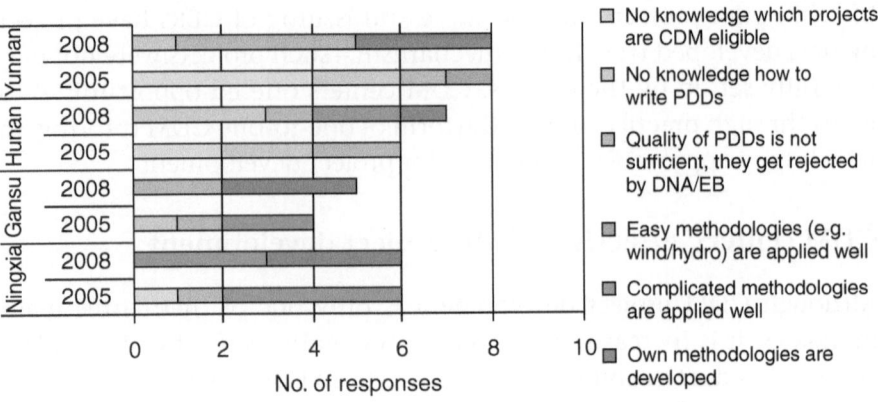

Figure 4.7 Assessment of the CDM capacity of CDM center staff

the carbon market, with Hunan and Yunnan doing worse than Gansu and Ningxia. This relation remained the same in the 2008 assessment, where doubts remained in Hunan and Yunnan about the quality of PDDs developed by the CDM centers. Not all interviewees were convinced by the CDM project-development capacity of the CDM centers. These doubts were expressed by two categories of respondents: a) project owners from Yunnan and Hunan who believed that the centers' experts could apply easy methodologies to develop, for example, hydropower projects, but would outsource more complicated project types to researchers; and b) local project developers, who raised this same point but also questioned the efficiency with which centers conduct their work.

Other factors that determine the effectiveness of CDM project development turned out to include timing and the availability of contacts with government officials and international CER buyers. The Ningxia CDM center was the most successful Chinese CDM center in terms of the proportion of projects it developed compared to the total number of CDM projects in its province. The crucial advantage of the Ningxia CDM center in this respect was its early positioning on the market. It was the first entity to spread information and provide training, and was thus able to establish itself as the pre-eminent CDM specialist in Ningxia.

Another important factor in successful CDM project development is the national and international outreach abilities of a CDM center. The centers in Ningxia, Gansu, and Hunan all have representative offices in Beijing, which ease access to CDM project development partners and potential CER buyers. The CDM centers also cooperate directly with foreign buyers, who are invited to the province for project sourcing, research, and project development. In the early market phase, this Sino-foreign cooperation for CDM project development consisted mainly of so-called 'CER selling conferences', at which contracts for joint CDM project development were signed. These conferences usually took place at the province's capital, and had the dual purpose of sealing cooperation projects and showcasing successful CDM projects to undecided provincial market participants. Among the four CDM centers under examination, the Hunan CDM center had the best international outreach. It had not only established a Sino-Swiss joint venture, but also had extensive bilateral business contacts with foreign companies – mainly buyers wishing to cooperate with the center on project sourcing. The center's director stressed that the center was on an equal footing with the foreign companies; it directly approached foreign companies for CER trading, but was also approached daily by those companies: 'Almost every day foreign companies come or phone to talk about new business' (Interview 56). In addition, the center's director participated frequently in international carbon events (visiting carbon expos, for example) and was a very active speaker at all kinds of national CDM-related events (Interview 27).

The Hunan CDM center was very active in establishing horizontal cooperation arrangements with other provincial CDM centers. It had already developed about ten CDM projects in other provinces (Interview 56). The center cooperates with several other provincial CDM centers. Its largest cooperation project, together with the Hubei CDM center, was the co-development of the Hubei provincial CDM market; both centers sourced CDM projects, and they cooperated in project development (Interview 57). The Hunan CDM center has also entered into an agreement with the Jiangxi CDM center to co-develop the provincial CDM market of that province.

But the four CDM centers under investigation differed in terms of the quantitative and qualitative outcomes of their project development. Performance indicators in this area were the numbers and types of projects that had reached either the validation or registration stage, or had been rejected. Figure 4.8 shows the differences between the four centers concerning the number of projects validated during each quarter: while the Hunan CDM center started relatively late, it showed a sharp rise in the numbers of PDDs developed, while the Ningxia and Gansu centers showed only very moderate increases after their initial level had been reached. The Yunnan CDM center had only three projects at validation in October 2009, although it was already in the process of developing four more projects at the end of 2007.

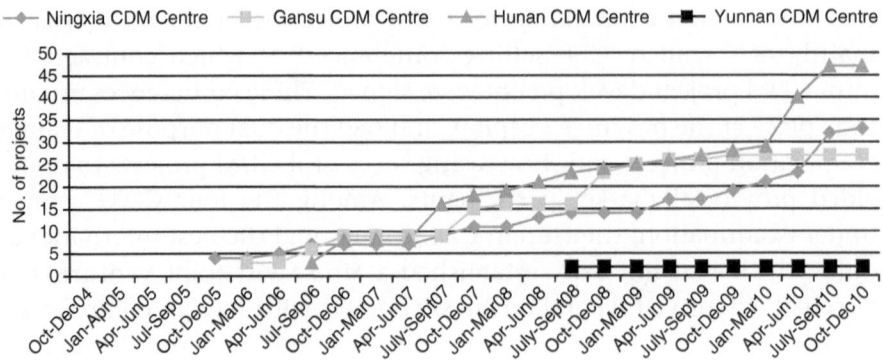

Figure 4.8　Number of projects undergoing validation developed by CDM centers
Source: Based on UNEP Risoe CDM/JI Pipeline Analysis and Database, 1 December 2010.

In order to assess the overall contribution of CDM centers to CDM project development in their respective provinces, Figures 4.9 to 4.12 display the development of each center's market share in its province. The tables show that the Ningxia CDM center was able to retain almost a monopoly of project development in its province, while the other three CDM centers were in fierce competition with private project developers. The figures reveal that CDM centers – with some exceptions, such as the Hunan and Ningxia centers – did not substantially alter market shares. In total, their market share was relatively small. Though it was considerable in some provinces, their overall impact on CDM market development in China must be seen as limited.

Among all CDM centers, the Hunan center has the largest number of CDM projects in the pipeline, and has also become the leading project developer in Hunan province (see Figure 4.11). By the end of 2010, over 50 of the Hunan CDM center's projects had been validated by DOEs, and 29 of those had been registered by the EB (UNEP Risoe, 2010). Ningxia provides a very clear illustration that CDM centers do not aim at increasing the ability of other provincial market players to develop CDM projects; instead, they try to retain their monopolistic position. The center, its private spin-off Beijing Keji, and its cooperating partner Shanghai Chuanji were thus able to develop all CDM projects in Ningxia until mid-June 2007 (see Figure 4.9). Project owners also confirm that they had no contacts with private consultancies, nor had they heard of any operating in Ningxia (Interviews 56 and 58). Only recently have CDM projects from Ningxia appeared in the pipeline that have been developed not by the center or its affiliates, but by other private project developers (see Figure 4.9). In comparison to the Ningxia CDM

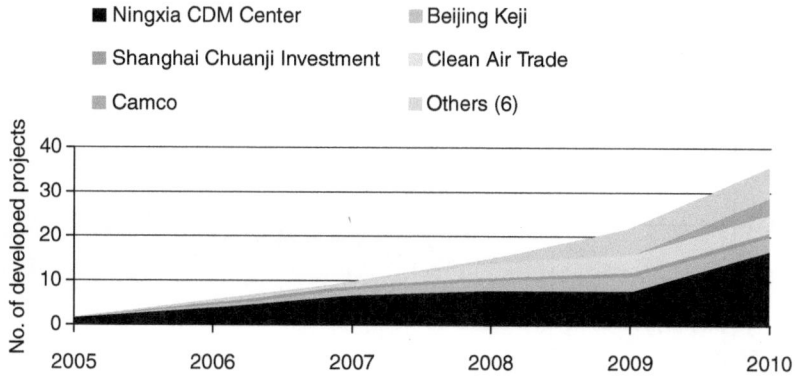

Figure 4.9 Ningxia CDM center's market share of projects

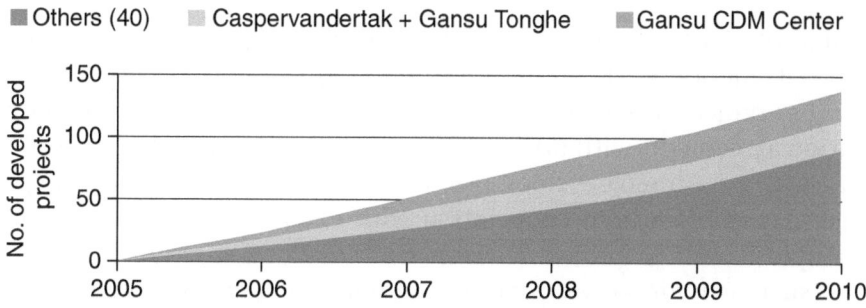

Figure 4.10 Gansu CDM center's market share of projects

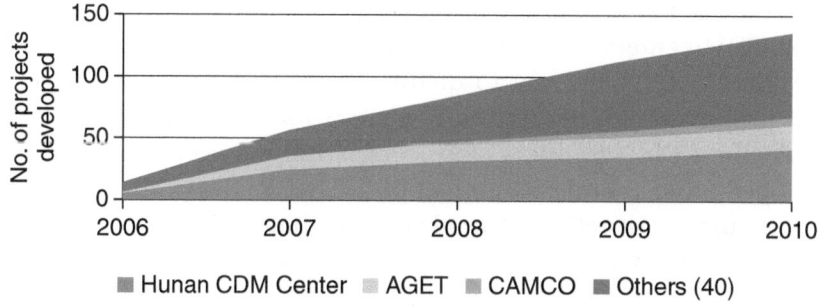

Figure 4.11 Hunan CDM center's market share of projects

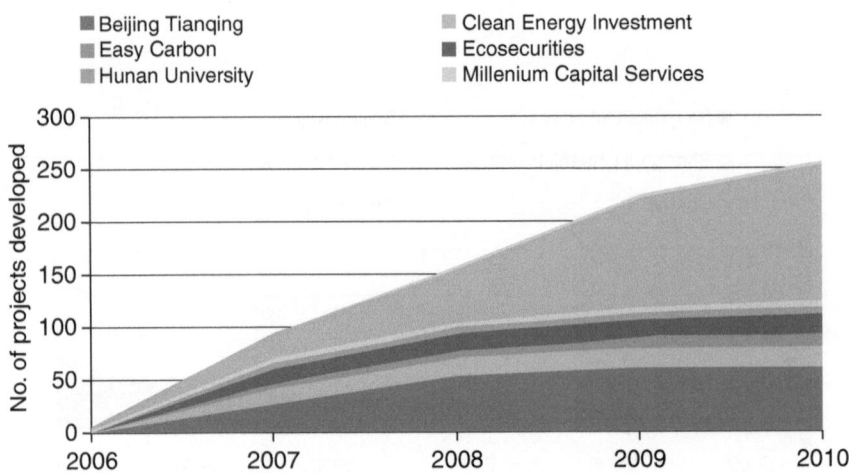

Figure 4.12   Yunnan CDM center's market share of projects

Source: Based on UNEP Risoe CDM/JI Pipeline Analysis and Database, December 1, 2010.

center, the Gansu center was unable (and seemingly not motivated) to retain its monopolistic position. Although its training was not targeted at other project developers, the Gansu CDM center had various cooperation agreements with other actors for joint project development. The Gansu CDM center develops PDDs with complicated methodologies for industry projects in cooperation with external organizations – for example, buyers or project developers. Projects with easy methodologies, such as hydropower projects, are developed by CDM center staff (Interview 49). The Yunnan CDM center has written more than 20 PINs and developed five PDDs for small hydropower projects, of which one received EB approval. The center had developed VER projects, but was at the time not able to find a buyer (Interview 64).

The CDM centers also showed some variation in the quality of the PDDs developed. One way to quantify that variation is by checking CDM projects developed by CDM centers for their registration status (see Figure 4.13). There is a close correlation between the number of developed projects and the number of rejected projects; this is probably quite natural, as it reflects the likelihood that a certain percentage of projects is always rejected. While the rejection rate of projects developed by the Gansu CDM center (8 percent) was slightly lower than the average Chinese rejection rate (8.4 percent), the other three CDM centers did worse than the average for project developers in China (Ningxia CDM center: 11 percent; Hunan CDM Center: 12.8 percent; Yunnan CDM Center: 33 percent) (UNEP Risoe, 2009). Interviewees, however,

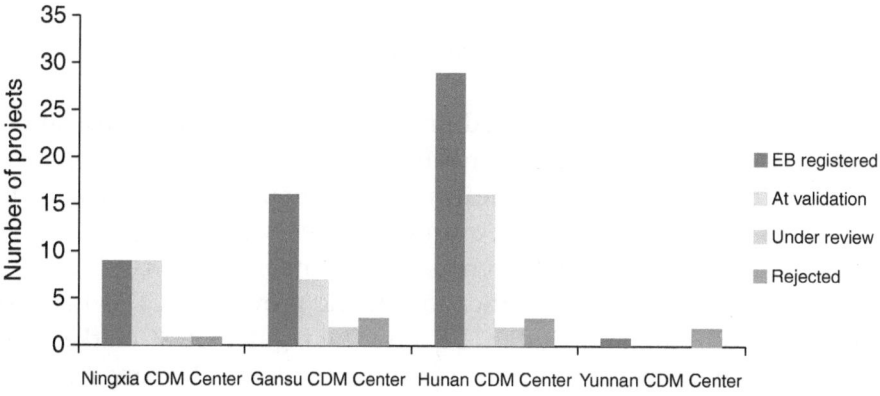

Figure 4.13   Comparison of CDM projects under review/rejected, EB registered, and at validation stage

Source: Based on UNEP Risoe CDM/JI Pipeline Analysis and Database, December 1, 2010.

also revealed some more subjective explanations for the high number of rejected projects in Hunan: apparently some of their PDDs were of lower quality, and had already been denied DNA approval.

Although the four CDM centers analyzed could not all live up to the project-development standards of international consultancies, they still constituted an alternative to the wild array of brokers that flooded the early Chinese CDM market. These individuals and companies were often incompetent, because they were not familiar with the CDM business as such; but they did offer their – sometimes dubious – personal networks for sourcing CDM projects from all over China. Due to the large size of the country and the importance of *guanxi*[1] in the Chinese business culture, having someone to introduce one company to another (especially a foreign one to a Chinese one) is very useful in sourcing projects (Interview 27). For this purpose, the semi-public CDM centers were a better alternative to the dubious private brokers, because they at least had some public support, and thus more legitimacy in the eyes of potential project owners.

CDM centers' impact on market diversification

The impact of the provincial CDM centers on the diversification of market actors was low. Training was targeted at government officials and potential project owners, but no other project developers. Although China's provincial CDM markets are diversifying in terms of private for-profit and non-profit organizations entering the market and acting as intermediaries and project developers, the increased diversification of market participants cannot be attributed to the activities of the CDM center.

Neither have CDM centers had much impact on product diversification, which in the CDM context refers to the methodology development for new CDM project types or the introduction of projects that go beyond the usual CDM approach – for example, projects generating voluntary emission reductions (VERs), such as Gold Standard projects. There is a high need for product diversification in the provincial CDM markets of Gansu, Yunnan, and Hunan, because the regular CDM market that focused on hydropower projects is coming to a close. For small hydropower, the potential in Gansu, for example, is diminishing, as possible power stations are already in the planning stage and have already signed their CDM contracts. But there would be considerable room to expand the CDM market to other sectors, such as energy efficiency, because the CDM is far more well-known in the renewable energy sector than in other sectors (Interview 18). Many interviewees expressed their opinion that there was still a need to introduce CDM methodologies that fit the local circumstances. Despite the need for new CDM methodologies, the CDM centers usually do not develop their own methodologies, because the methodologies for their most common project types – wind and hydropower – are already well used, and because developing a new methodology would require time and good English skills (Interview 41). Although some interviewees mentioned that international information exchange on new methodologies (the use of solar cookers, for example) would be very useful, there is no such exchange taking place between project developers; they instead refer to the PDDs from foreign projects that are available on the internet (Interview 47). The danger of having innovative PDDs copied as soon as they become publicly available on the internet was also raised by two project owners as a limitation for methodology development in general (Interview 42; Interview 47). One therefore has to judge their impact on product diversification within the provincial CDM markets as minimal.

Barriers to full participation in the carbon market still remain. A representative of Ningxia's Science and Technology Department states that Ningxia does not have enough qualified people, especially young people. With many older leaders retiring soon, he would prefer to have 'a few young PhDs to continue the CDM work.' The lack of a qualified workforce was also raised by the director of the CDM center as a limiting factor for Ningxia's CDM market. Another representative of the CDM center pointed out that an information deficit as to local situations makes international activities like those of the CDM very difficult (Interview 42). A third barrier to CDM diffusion is the unwillingness of many companies to take on any upfront investment for CDM projects because they still perceive these to bear a high risk (Interview 43). Project owners complained that the CDM does not contribute to achieving the financial closure of projects because CER revenues only come in after project implementation. The CDM is thus only a finishing touch, and not a substantial help in times of need. A fourth fundamental barrier to CDM project development in Ningxia is the continuing reluctance of banks to take CDM revenues as collateral. Finally, several interviewees complained that the CDM had so far not been sufficiently able to deliver on technology transfer. Its greatest contribution so far was the inflow of foreign capital, but – probably also because of the advanced stage of hydropower technology in China – it had not brought additional technologies (Interview 61). The director of the Hunan CDM center saw the main reason for the lack of technology transfer as foreign companies' focus on CER trading, and their lack of interest in technology transfer. In his opinion, one measure that might foster technology transfer would be increased regulation by the EB of how technology transfer should take place in CDM projects. Another option would be to have technology transfer included as a component of the ERPA (Interview 56).

A number of challenges must be met in relation to climate change: the companies' capacity for environmental management needs to be increased; rule implementation and the monitoring capacities of the government need to be strengthened; and public awareness of climate change must be raised through media campaigns (Interview 45). A need remains for more CDM capacity development – especially in the registration process, where there are still many difficulties, as well as points people do not understand – as well as in new methodologies, which change all the time (Interview 46). Although there are plenty of CDM training opportunities available, for example through the CDM training center of the NDRC (Interview 48), the rising interest in products other than the CDM – for example, VER projects – has not yet been met with adequate learning opportunities (Interview 48).

Summary of each CDM center's performance

The analysis above reveals that the Ningxia CDM center has made a substantial contribution in spreading information on the CDM in the early provincial market. Its training, which was mainly targeted at government officials, also contributed to creating a CDM-friendly attitude on the part of the provincial government, which took shape in several supportive policies. The CDM center has also proved very successful in CDM project development, and was able to enjoy a monopoly in the market until October 2007. The center took a facilitating role in CDM market development at first, but later concentrated on its own profit-making activities. After the closure of the Sino-Canadian capacity development program it refrained from offering additional CDM training, and it is now not supporting, but instead preventing other project developers from entering the Ningxia market. The center's training was very successful in reaching local government officials, who supported the CDM through various political measures. But the center was not able to reach out to financial institutions, and the lack of upfront financing remains the biggest barrier to CDM project development on the Ningxia CDM market.

The Gansu CDM center was the first CDM market actor in the province. This position of 'early bird' gave the center a window of opportunity that it was able to use effectively. It contributed substantially to raising CDM awareness and capacity among market actors – although only in relation to government officials and potential project owners, rather than other project developers. The Gansu CDM market soon became highly competitive, with one local and one international consultancy becoming active. Facing competition, and being involved in many of the non-CDM activities of its provincial government, the Gansu CDM center has not been quantitatively strong in CDM project development, although a relatively large number of its projects are eventually registered. The center represents a good example of how public status is advantageous in the early phases of market development, but cannot maintain a dominant market position in competition with private consultancies. The center does not expect the CDM to be its core business function in the future; it will also have to take on other research and technology-promotion tasks for the government.

The Hunan CDM center can be identified as an aggressive newcomer to the CDM market. Despite its relatively late establishment, the center was able to establish itself as one of the province's main project developers. The Hunan center is a private company that is only loosely attached to the provincial government. It has limited political support, but was part of the Sino-Canadian CDM capacity-development program. Its information-dissemination activities and training only took place within this program, so its impact on CDM awareness-raising and capacity development are only moderate. But the center has performed very well in CDM project development, thanks to its aggressive marketing strategy, its motivated staff, and its many national and international outreach activities. The Hunan CDM center is thus a good example of how well a private company with government support can do in CDM diffusion; but it also shows that activities promoting more public goods, such as awareness-raising and capacity development activities, tend to be neglected if they are not closely linked to the profitable PDD development activities.

The Yunnan CDM center can be described as a 'laggard' on the already-flourishing Yunnan CDM market. In contrast to other CDM centers, the Yunnan center is only a part of the local government, and has neither the management structure of a private company nor a private spin-off component. In addition, the center was established relatively late, at a point when the Yunnan CDM market was already well-developed, information was spread, and CDM capacity established, so that most market actors concentrated on CDM project development. Despite its mandate for CDM capacity development, the Yunnan center operated almost no CDM-promotion activities, and conducted only a few CDM training programs. Impacts in these two fields have been low because of the limited scope of their activities, and because of their entry into the market occurring when all hydropower companies, at least, were already very familiar with the CDM. The late entry of the CDM center and its perceived low quality of services are reasons why the center has also been unsuccessful in CDM project development. It remains to be seen whether the CDM center will be able to establish its position in a market that is already highly competitive – not only a result of the many private consultancies that are active in Yunnan, but also to the existence of at least one other semi-governmental organization that claims the title of 'Yunnan CDM center.' The government affiliation of the Yunnan CDM center has not proved at all advantageous, because the government itself does not follow a coherent CDM support strategy, and because government attachment seems to be a barrier in the competitive market phase of CDM diffusion and market diversification.

5
The Role of Diffusion Catalysts

CDM centers' varying degrees of effectiveness

The four case studies revealed a variation in the effectiveness of the four CDM centers under examination. As the 'effectiveness' of the CDM centers refers to their ability to achieve their objectives, different kinds of outputs, outcomes, and impacts need to be considered when evaluating it. For the four CDM centers under consideration, the analysis focused on three intervention strategies: CDM awareness-creation, CDM capacity development, and CDM project development. As far as their direct impact on CDM diffusion is concerned, the effectiveness of the CDM centers varies considerably. Although the intervention strategies of the CDM centers were of the same type, they had different impacts on each diffusion phase, depending on the scale and quality of their activities.

The CDM centers' information-dissemination strategies included setting up websites, publishing CDM brochures and newspaper articles, and facilitating radio and television features. CDM centers could also use government channels to inform project owners about the CDM and invite them to CDM conferences. The bulk of these CDM-promotion activities took place in the early market phases. These interventions contributed to an increase in CDM awareness among market participants: impacts were high in Ningxia and Gansu, where the CDM centers were among the first sources of information on the CDM. Their impact on awareness-raising can be straightforwardly acknowledged, because in that early phase of CDM market development in China, hardly any other national or local information sources were available. On the other hand, the impact of the Hunan CDM center on CDM awareness-raising must be considered to be moderate, and the impact of the Yunnan center low, because these two centers started to disseminate CDM information at a point when several other national information sources were already available – for example, private consultancies were promoting the CDM at the provincial level. Due to an aggressive marketing strategy and some innovative features of their website, the Hunan CDM center was nevertheless able to establish itself as a trusted CDM information source in the province, while interviewees were barely aware of the Yunnan CDM center's information activities.

The CDM centers' capacity-development strategies included CDM training sessions and workshops that took place at the provincial capital, and sometimes also on the district and country level. CDM training and workshops mainly targeted project owners and government officials; journalists sometimes participated, though financial institutions were not interested, and other project developers were not invited. Some CDM centers only conducted the essential training that was required by the Sino-foreign capacity-development programs. Other CDM centers also conducted additional CDM training at the sub-provincial level, or targeted new sectors, thus trying to reach out to new PDD customers. These interventions helped project owners to understand the CDM requirements and judge their project's CDM eligibility, and increased their ability to use the CDM as leverage in loan negotiations. Training and conferences also helped to create CDM capacity among local government officials, which they sometimes used to draft CDM-supportive policies. Most of the CDM centers also used their standing as sub-departments or government-affiliated entities, using the official communication channels within a provincial government to provide CDM information and policy suggestions to their superiors. Besides personal interaction, this took the form of written reports. Officials from the provincial government thus often supported the center's work, lifted the CDM on the political agenda, and eventually drafted notes, reports, and policies supporting CDM project development in their province. This in turn had a positive effect on project owners' trust in the CDM, as they generally understand government support for an initiative as a sign of its trustworthiness. Importantly, the capacity-development measures thus had increased trust in an international mechanism that had initially been regarded as a 'cake falling from the sky,' which nobody would dare to believe in. Concerning the overall impact of the centers' capacity-development strategies (see Table 5.1), none of the centers can be assessed as having a high impact, because all omitted one important group from their capacity-development activities: project developers. Instead, activities only targeted potential project owners and government officials (and financial institutions, which showed hardly interest in participating). Similarly, as with information dissemination, the situation in Ningxia and Gansu was conducive to the CDM centers having a directly traceable impact on CDM capacity development, because they were among the only entities providing training and workshops in the early CDM years in China. Their impact can nevertheless only be assessed as moderate, because they more or less stopped their capacity-development activities with the phasing out of the Sino-foreign assistance programs. The Hunan CDM center, on the other hand, picked up capacity development measures at a relatively late point in time (when other entities were also offering such activities), but continued to offer CDM training even after the phasing out of the Sino-foreign assistance programs. The Yunnan CDM center had so far only organized a few training sessions, but planned to enter new sectors and a more local level of operation with its capacity-development activities in the future. Based on the data available for this book, however, their impact on CDM capacity development must be judged to be low.

All CDM centers are still very active in CDM project development; they offer PIN and PDD development and match-making with potential buyers, and they participate in ERPA negotiations. Most of the CDM centers also actively advise their provincial government on CDM policies and other climate-related matters – the Hunan CDM center, for example, is involved in developing the province's climate change program. These interventions had a verifiable impact on diffusing the CDM by offering direct CDM project-development services. While there was only a market monopoly on the part of the center in Ningxia, there are some other CDM centers that established themselves as successful project developers. Because figures are available on the share of the CDM centers in their provinces' successfully developed projects, their impact on CDM diffusion in terms of the number of CDM projects developed can be clearly assessed. Although the Ningxia CDM center lost its monopolist position on its provincial market in 2007, in December 2010 it is still responsible for 56 percent of the CDM projects that reached validation stage in Ningxia, and its impact on diffusion can therefore be considered to be high. An outstanding characteristic of the Ningxia CDM center is thus its status as an 'early monopolist.' The same assessment goes for the Hunan CDM center, which used to be responsible for over 50 percent of the projects developed, and in December 2010 still had a 34 percent share of projects in the province. Then Hunan CDM center can thus be labeled the 'aggressive newcomer' to the market (see Table 5.1). Although the Gansu CDM center still retains a considerable share of the CDM projects in the pipeline of its province (19 percent in

December 2010), its impact on CDM diffusion can only be assessed as moderate, because it faces heavy competition from at least one Gansu-based project developer with comparable project-development rates. The initial success of the Gansu CDM center can thus be summarized as being due to its position as an 'early mover on a competitive market' (see Table 5.1). The Yunnan CDM center must be assessed to have only a low impact on CDM diffusion, because its share of projects in the pipeline makes up only 1 percent, and because analysis reveals that 'turfs' in Yunnan have been clearly divided up by the private consultancies, leaving little scope for future improvement in the center's share. The Yunnan CDM center could thus be classified as a 'laggard' in comparison to other CDM centers (see Table 5.1).

In terms of their indirect impact on market development, the case studies revealed that none of the CDM centers examined had a systematic approach towards diversifying the market – a task that seems not to be seen as part of their mandate. The CDM centers are not very active in market-diversification activities. CDM centers do not engage in CDM methodology development, but they have begun to include information on the VER market in their publications, websites, and training programs. Although it was not clearly intended to be the case, CDM centers have some influence on CER seller and purchaser demographics. Due to the currently demand-driven nature of the CER market in China, CDM centers have a relative large choice of buyers for their CERs, and are able to increase their choice by attending international conferences (for example, carbon expos) and by setting up sub-branches in Beijing, where they are closer to the buyers. They can also, in turn, increase their choice of project owner, because some of them now turn to other

Table 5.1 Summary of CDM centers' impact during different phases of CDM diffusion and market development

	Ningxia	Gansu	Hunan	Yunnan
Awareness raising	High	High	Medium	Low
Capacity development	Medium	Medium	Medium	Low
CDM adoption	High	Medium	High	Low
Market diversification	Low	Low	Medium	Low
Type of center	Early monopolist	Early mover in competitive market	Aggressive new-comer	Laggard

provinces to source new CDM projects – the Hunan CDM center, for example, has a cooperation agreement with the Anhui CDM center to develop projects jointly in Anhui province.

CDM centers' overall contribution to market development

My analysis so far has explored how the differences between the CDM centers in their contribution to market development can be explained for the three intervention areas. But a note of caution has to be struck on the overall relevance of CDM centers for CDM market development in China, because, in all three intervention areas – most obviously in project development – CDM centers face strong competition from private business actors, other market actors, and even from the NDRC itself. Since a multitude of diverse actors is involved in the CDM market, it is also not surprising to see other actors than the CDM centers taking up activities that contribute to CDM diffusion and market development, even though their motives are profit-oriented (a tendency observed for capacity development processes in general in, for example, Fukuda-Parr et al., 2002: 20). While many interviewees mentioned other actors besides the CDM centers as important in awareness-raising and capacity-development, their impact in comparison to that of the CDM centers is hard to quantify.

The impact of private consultancies in project-development is much easier to grasp, thanks to the plentiful data on CDM project developers. Compared to the total number of Chinese CDM projects, the overall contribution of the CDM centers to project development is modest in comparison to that of private consultancies (see Figure 5.1). This comparison reveals that even very strongly performing CDM centers

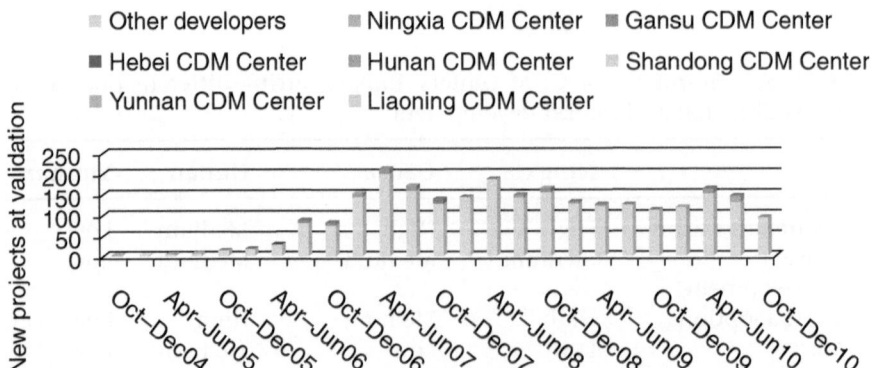

Figure 5.1 CDM centers' projects at validation as a share of total projects in China

Source: Based on UNEP Risoe CDM/JI Pipeline Analysis and Database, December 1, 2010.

face strong competition from private consultancies, which might ultimately turn out to be the better choice for facilitating local CDM markets. At least one representative of the Chinese central government also expressed his conviction that the high number of CDM projects in China is due to the activities of private consultancies, which also would have contributed to capacity-development (Interview 1).

If one compares the overall project-development data for the Chinese provinces, the similar rates of CDM projects in the pipeline for provinces with comparable CDM potential are striking. Although the CDM centers show a high variation in their PDD development, their provincial markets nevertheless show similar degrees of market development. It seems that, in provinces in which the CDM center is not performing strongly, such as Yunnan, private consultancies had already taken up a substantial share of the market at an early stage, thereby contributing to CDM diffusion.

As far as CDM awareness-creation and capacity-development are concerned, the attempt to identify relevant for the effectiveness of CDM centers at the local level highlights the implicit paradox that it is not primarily local factors, but rather international ones that influence local awareness and capacity. This does not apply only to the CDM market, but tends to become the norm in the process of globalization. For our purposes, the fundamental question is thus what constitutes a local phenomenon, as opposed to the spill-over effect of a global process? Applying it to the question of what factors stimulate learning and innovation, Malmberg and Power call this situation an implicit paradox (Malmberg and Power, 2005: 273). How can one identify the causes of local learning and innovation processes when these are at the same time embedded in globalization, with its increasing prevalence of long-distance flows not only of goods and money but also of market signals and information. In the processes of learning and innovation, then, which factors are localized and which are not?

The case studies revealed that many international, national, and local sources of information and learning opportunities have impacts on provincial market players' CDM awareness and capacities. One major international source of information is the UNCCC website, which was ranked by interviewees as the most frequently visited website in this study. Nationally, the activities of the Chinese government itself can be seen as alternative explanations for local CDM market development. Since the initiation of the Chinese CDM market, the central government has provided information campaigns and capacity-development measures (often in cooperation with foreign donors, but increasingly by itself and for profit). Many interviewees stated that it was the Chinese DNA's website that they turned to first for CDM-related information. This was explained by the 'consolidated information' it offered, which could be trusted for its accuracy. Once more, a government institution seemed to receive a higher level of trust for its CDM information than did private companies. The national-level 'CDM center,' a training institution attached to the NDRC, offers for-profit CDM training targeting potential project owners, project developers, and local political leaders from all over China. International development agencies also initiate CDM awareness, creating campaigns that are not directly linked to the provincial CDM centers. For example, the European Commission has organized 'CDM road shows' in several provinces, and German International Cooperation (GIZ) has conducted CDM training in several provinces. Because these are usually one-off events that do not inaugurate a local institution like the provincial CDM centers, and because interviewees did not mention these events as sources of information or capacity, their impacts are not considered in this study.

Despite the expectations of innovation theory, as well as previous experience with the development of new markets, it was not only public but also private actors that became active in awareness-creation for the CDM in various Chinese provinces. The main incentive for private project developers to become involved CDM awareness-creation in the provinces is that of sourcing new CDM projects. They are involved in heavy competition to reach potential project owners first, and to convince them of the feasibility and profitability of turning their projects into CDM projects. Their strategies for awareness-creation differ from those employed by the CDM centers. As they cannot rely on government channels for reaching potential project owners and distributing information, they have to establish their own private networks. Three possibilities were mentioned in the interviews on how to make use of local networks for CDM awareness-creation. One option is to cooperate with already existing provincial-level consultancies that have their own networks. The other is to set up provincial-level offices which then establish their own network. Provincial offices often carry out project-sourcing alone, while at least the development of complicated types of CDM project is carried out by the consultancies' experts in Beijing. As well as for project-sourcing, it has proved helpful to be close to project owners in order to supervise them in the whole process of project-development. Having a provincial office can help to save on travel costs and time, and enables staff to maintain close contact with project owners and government officials, as well as to be up-to-date on local information. Another option mentioned by an interviewee was that of using the personal networks of the founders of private consultancies, because those founders tend to be well connected in their province of origin, or in the sector of their previous profession.

In addition, project owners and industry associations also disseminate information vertically within their organizational structures in order to inform their colleagues at the provincial level about the possibility of running CDM projects. For example, a Ningxia branch of one of China's five big power companies received CDM information and training organized by its headquarters. Hydropower industry associations are also active in providing CDM information to their members; training is not held, but members informally exchange information about project developers and their successfully developed CDM projects.

There are some instances of CDM awareness-creation campaigns by NGOs. The Beijing-based NGO, Global Environmental Institute (GEI), has created a CDM project shortlist for its EU-funded project, 'Capacity Building on Business Opportunities for CDM Projects in China.' The international NGO, The Nature Conservancy (TNC) is lobbying for the inclusion of 'Climate, Community and Biodiversity' (CCB) standards in CDM projects in Yunnan. The local NGO, Green Camel Bell, is cooperating with a local community and a CDM project owner in Gansu for the establishment of a small-scale biogas CDM project.

An alternative explanation for an increase in CDM capacity in the four provinces is the training conducted by private market actors. Private actors involved in CDM capacity-development programs include project developers and DOEs conducting CDM workshops and training in the provinces targeted at sourcing CDM projects and promoting their company and their validation and verification services. Also, several DOEs, including DNV and TÜV Süd, have conduced capacity-development training programs on the CDM in its early market phases and for provinces and sectors that were lagging behind. DOEs are also active in providing training on how to monitor CDM projects (Interview 27). While these contribute to capacity-development, the first type of training was conducted as promotional activity to make the DOEs known to potential customers, while the latter type have been conducted to enable project owners to implement their projects effectively and monitor emission reductions. This will eventually ease the verifications that the DOEs must carry out. Apparently the Chinese government has also outsourced capacity-development to private companies. For example, the NDRC has entered an agreement with the CDM consultancy Mitsubishi UFJ to train provincial and local officials on CDM-related issues (GTZ, 2008).

Although CDM centers may perform well in terms of their outputs – their intervention activities – they do not necessarily have a strong impact on CDM diffusion and market development. Provincial CDM centers have had an impact in launching and consolidating their provincial CDM markets, but their impact on maturing markets by diversifying these is low. The centers' focus on providing information to project owners and the exclusion of project developers from CDM training lead to the conclusion that CDM centers act not as market-facilitating institutions, but mainly as CDM project developers. This tendency is confirmed by statements of CDM centers' representatives, who see the expansion of their PDD development services as their future business development strategy. While these tendencies are not in line with theoretical models of diffusion catalysts, they are in line with the objectives of the donors' capacity-development projects, which also focus mainly on the development of PDDs and the generation of CERs. On the other hand, private actors – for example, project developers, buyers, and even DOEs – have launched their own CDM training programs in the provinces. Although they do this with the goal of reaching out to more potential project owners, they thereby contribute to provincial capacity development. This observation contrasts with the theoretical assumption that public actors are chiefly responsible for capacity development in nascent markets.

But even if some of the CDM centers' services are of good quality, they face heavy competition from public and private actors that either offer services earlier, or offer services of even better quality, scope, and speed. Table 5.2 summarizes who are the main players in various activities contributing to CDM diffusion and market development, both at the provincial and national levels in China, and at the international level.

Table 5.2 Who is involved in what types of activity in CDM market development

Area Level	Awareness-raising	Capacity development	Project development
International	UNFCCC, researchers	–	–
National	NDRC, donor agencies, consultancies, DOEs, NGOs	NDRC, donor agencies	consultancies
Provincial	CDM centers, consultancies, DOEs	CDM centers, consultancies, DOEs, NGOs	consultancies, CDM centers, researchers

The impacts of organisational structure and the timing of interventions

This section deals in detail with selected issues emerging from the empirical analysis, because these issues were found to be decisive in explaining the effectiveness of the centers. They include the advantages and disadvantages of being a semi-public agency, and the timing of interventions. This section also aims to explain why these two issues proved so important. Explanations rely partly on factors identified in the empirical research, but also on linkages to assumptions from innovation theory (see Chapter 4). I thereby offer an assessment of the extent to which my empirical findings corroborate theoretical assumptions, and identify the shortcomings of innovation theory in explaining several important aspects of those findings.

Advantages and disadvantages of being a (semi-)public agency

The ownership structure of a CDM center has been shown to be important for its effectiveness. The nature of CDM centers is hard to pin down. CDM centers exist somewhere in the space between private and public, and they can be both. The Ningxia and the Hunan CDM centers each consist of both private and public companies, which have different names and different corporate identities, but have the same director and the same staff. Their character also varies in terms of location and in time. In some cases, these centers are merely part of the local government; in others, they are private companies in which the state owns shares. With time, they tend to move from being public to becoming private. CDM centers can thus be viewed as one form of hybrid actor. Because this phenomenon has become so common in contemporary China, Chinese interviewees alluded to a Chinese saying describing this situation: 'one person with two heads.' This situation is also common in many other industry sectors (Solinger, 1992: 135).

There are three advantages to being part of, or at least affiliated with, the provincial government: the opportunity to use the government's system of communication; reliable financial and political support from the government; and the possibility of benefiting from the government's reputation.

Communication channels

Using government channels to spread CDM information horizontally to different market actors, and vertically to different provincial levels, was one advantage of government affiliation that could be observed for all four CDM centers examined (for a good overview of how cadre networks provide information to local enterprises in general, see Oi, 1994):

> If a CDM center has a government background it will spread CDM knowledge more efficiently. As a government organization, it enjoys a high level of credibility, and can use its networks with district and county DRCs. If it has a government background, it can initiate CDM projects earlier and more successfully. (Interview 69)

This advantage is confirmed by a representative of the NDRC:

> [Centers] receive political support, are close to officials, and take part in information flows. Their attachment to the local government is an advantage, because they take part in the internal information flow. (Interview 1)

Financial and political support

Being partly on the government's payroll places CDM centers in the very comfortable situation of being able to pursue lucrative CDM-related activities when prospects are good, but also leaves them the option of withdrawing to a position of simply being an entity working for the government, perhaps making less profit, but securing their existence:

> An advantage for CDM centers in being a part of the government is that their financing is secure. Our Ningxia CDM center is a private company with government support; it can operate independently. If the CDM market is good, it can develop projects; if the CDM market is not good, it can become a consultancy for the government, at least helping the government and companies to develop a good regulatory regime– or even better, to do policy research. (Interview 43)

Enhanced reputation

Government affiliation also increases the credibility of CDM centers. Interviewees explained why government affiliation is so important: in a situation of almost uncontrolled economic growth, a government affiliation acts as a guarantee that an entity is not just a flash in the pan, but will continue to exist and honor any liabilities. The ability to trust an organization seems fundamental for project owners when they are deciding who to cooperate with. The tendency of project owners to trust government institutions more than private companies is thus one more explanation for the better performance of CDM centers with regard to project development compared with their private competitors in the early market phases. The reliance on CDM centers for CDM project development can be explained by the fear on the part of many project owners of becoming overburdened by the complex bureaucratic requirements of the CDM (this reason was always given first when interviewees were asked about the CDM's disadvantages); these requirements are not the focus of their business, and they were unwilling or able to invest much time or effort in them. One reading of such responses is that project owners were neither willing nor able to make fully informed decisions about the best choice of project developer, and thus relied on an emotional decision. As they seemed to be limited in their ability to gather and process all relevant information, they were acting within the constraints of a 'bounded rationality' (Simon, 1972). At the same time, there was also evidence of mistrust on the part of project owners of private project developers, as one international project developer explained:

> One difficulty that has been mentioned for Gansu, but which is probably also given for other places, is that the role of project developers is sometimes not clearly understood by project owners, who think that buyers are their main partners and are then not very transparent with the intermediaries. (Interview 52)

These findings on the importance of the trustworthiness of CDM centers are supported by experience with BDS centers and by research of new institutional economists on the issue of trust in contractual relationships (Keefer and Knack, 2005). In his comprehensive analysis of international experiences with BDS centers for small- and medium-sized enterprises, Jacob Levitsky also observed an initial mistrust on the part of these companies towards consultancies offering BDS services. Only time and the work of private BDS suppliers increased companies' trust in these consultancies and willingness to cooperate (Levitsky, 2000: 11). Some authors with a background in new institutional economics see a clear hierarchy of importance between trustworthiness and expertise in terms of their impact on the willingness of business partners to cooperate. In the absence of reliable knowledge about someone's expertise, trust becomes the paramount factor in a decision on whether to cooperate with them. Trust can be defined as 'the mutual confidence that no party to an exchange will exploit another's vulnerabilities' (Sabel, 1993: 1,133). Trustworthiness in this context refers not to trust in a relationship between two actors, but to a characteristic feature of one actor (Barney and Hansen, 1994: 176). Because the early CDM market in China was characterized by highly asymmetric information, with the majority of CDM experts, and of people in general, not believing that the CDM in principle could be trustworthy, no local company would have entered a CDM project development contract if it had not whole-heartedly trusted its counterpart. This was even more true because no normal company would have had the capacity to check on whether the services offered by a project developer were of good quality until the moment when a CDM project was applying for registration. Choosing a project developer was thus not an easy task, especially in a country where private companies have often been suspected of opportunistic behavior (Pearson, 1997: 59). Among others, this mistrust of private CDM brokers (especially international ones) was one major reason why the Chinese central government decided to set up government-affiliated CDM cent-ers in the remote provinces. Government affiliation should also have helped the CDM centers to gain bargaining power vis-à-vis interna-tional companies, because these would need to abide by government decisions (a general advantage of public service units already observed in Nee, 1992: 5). Even today, representatives of international consultan-cies feel that the Chinese government favours Chinese companies and research institutions for CDM project development (Interview 18).

The need for credibility and trustworthiness was especially strong in the early market phases, when companies did not believe the incentives resulting from the CDM to be real. In later market phases the impor-tance of a government affiliation decreased, because companies tended to prefer private consultancies who by then could often establish cred-ibility by their successful project-development record (Interview 69). If the CDM center cannot offer a record of such successes – either because it has in fact been unsuccessful or because it was established late – it is perceived to be less reputable, regardless of any government affiliation, than a private consultancy.

There also several downsides to being part of government, or affiliated with it. They include potentially incompetent staff with bureaucratic attitudes, many additional government obligations, and less freedom in management decisions.

Less well-qualified staff

As the case studies also revealed, being part of the government often implies that a CDM center has to accept transferred staff from other gov-ernment departments, who are usually not CDM experts. Research on other types of public service unit has even indicated a certain 'garbage can' character of such agencies, as these are sometimes forced to take on redundant staff from their superior organizations, which have to reduce their official staffing levels. This procedure affects overall staff quality, and in turn the institutional capacity of the public service units (Lo et al., 2001b: 59). This 'garbage can' tendency of public service units could not be confirmed for the CDM centers, whose representatives claimed to have well-qualified and able staff. However, some interviewees from private consultancies had a more critical view of the centers' staff, and remarked that their secure lifetime position could also make them less motivated and more lax in relation to their work assignments, whereas a job at a private company would always demand effectiveness: 'In the government it is much more relaxed – you go for drinks, you have a three-hour lunch break, drink; so I guess it is more difficult to be efficient in the afternoon' (Interview 52). Other interviewees voiced the opinion that having a more bureaucratic attitude towards job performance would induce CDM centers' staff to work more according to commands from above, with less service and customer orientation:

> The main difference between public and private entities is that the government works by command. They tell a project owner to attend a training program and to start a CDM project, but nobody cares or knows how to bring a project to success. Government entities have their leaders, but a consultancy is offering services; they listen to what the project owner wants and needs. So government entities are quick to source projects, but not always successful in implementation. (Interview 70)

These kinds of non-performance-based attitudes were also confirmed by a representative of the Ningxia Development and Reform Commission, who remarked that staff in local government generally worked less efficiently because, for them, 'it doesn't matter whether they develop one CDM project more or less, and they are also not very innovative regarding technologies' (Interview 43). One consequence of such deficits in staff qualifications and motivation is that some CDM centers need to subcontract PDD development either to private consultancies or to research institutions (Interview 1). On the other hand, CDM centers can be an attractive employer, because some of them offer better-paid performance-based job contracts (for example, Hunan). This comparative attractiveness as an employer has already been acknowledged for other types of public-service units in China. For example, Lo et al. observe that it is beneficial for cadres to switch to a

merit-based (and better paid) job in public service units before return-
ing to the bureaucratic structures of local Environmental Protection
Bureaus for the benefit of their political career (Lo et al., 2001b: 54).

Less freedom in management decisions

Being part of the government also ties CDM centers and their staff into
various intra-government procedures and obligations. On one hand,
this involvement can be used for internal lobbying, or at least for sharing important information (in the form of reports, for example). On
the other hand, much time and resources must be spent on meetings
and conferences, and on fulfilling non-CDM-related tasks delegated to
the CDM center. In particular, the heavy workload generated by writing reports to superior departments, sometimes on a monthly basis,
is a relic of the Maoist period (Oi, 1995: 1,145). This situation has, for
example, led to negative assessments of the Yunnan CDM center by
some market participants, who perceive it simply as a 'talking shop':
'I've never heard about any of their activities. They never do anything;
they only participate in the meeting held by the NDRC' (Interview 69).
A lack of enthusiasm and efficiency is thus identified in the work attitude of CDM center staff, in contrast to that of private consultancies,
which need to deliver high-quality work:

> The [Yunnan] CDM center treats CDM projects as only one of its
> functions as a government institution. The private companies treat
> the CDM as their business. They want to get profit from it, so they
> try their best. The governmental CDM center is not enthusiastic.
> (Interview 69)

Some CDM centers must also deliver various non-CDM-related services to the government. This limitation of their CDM work is acknowledged by a representative of the NDRC: '[Government affiliation] is a
disadvantage, because they have to follow their leaders' instructions;
they have a lot of bureaucracy to deal with and are thus not very flexible' (Interview 1).

A continuing dependency of public service units on their government
patrons has also been observed by Oi, who remarks:

> The flow of ideas is now two-way; individual enterprises are free to
> do their own market research and development of product lines.
> But local governments remain important for facilitating the actual
> implementation of these ideas, regardless of the source. Local cadres

use their expansive connections and bureaucratic position to secure information that will serve local economic growth, particularly as China enters the more competitive international market. (Oi, 1995: 1,140)

Table 5.3 summarizes the advantages and disadvantages for the CDM centers of having either a public or private structure. Please note that 'public' and 'private' are only labels used for analytical purposes, whereas in reality the organizational structure of CDM centers is in flux, and they take different positions on the public–private continuum.

Naturally, the interviewees were divided in their assessment of the overall effectiveness of CDM centers and the usefulness of their government affiliation. Private consultancies, as potential competitors of the CDM centers, expressed negative evaluations, while government

Table 5.3 Advantages and disadvantages of private or public structure for CDM project development

Advantages	Disadvantages
Public institutions	
• Use of government networks to reach out to potential project owners down to the township level	• Low motivation and low efficiency
	• Research obligations
	• Bureaucracy
• Participation in internal information flows	• Diverging political objectives, bound by leaders' instructions
• Trustworthiness	• Work by command; no customer orientation
• Political support	
• Reduced staff fluctuation due to secure employee position	• Less-qualified staff
	• Less contacts with buyers
• Secure financing	
• Local presence	
Private companies	
• Efficiency and higher motivation	• No official support
• Higher salaries; better-qualified staff; ability to source staff from all over China	• No use of government communication channels
	• Need local brokers to introduce consultancy to potential project owners
• Ability to offer better prices and services	
• Better contacts with investors, especially if Beijing-based	
• Ability to make quick decisions in a high-risk business	

representatives – as supporters of the CDM centers – shared a more positive assessment. In essence, the advantage of being closely affiliated with the government is that it inspires greater trust and legitimacy in the eyes of potential project owners, and the disadvantage lies in being regarded as less professional and efficient. Representatives of private consultancies share the most common perspective: they acknowledge the important role of CDM centers in CDM promotion and capacity development, and they even admit their competitiveness in CDM project sourcing. They do not, however, see the CDM centers as equally competent agents of CDM project development, and demand that CDM centers refrain from offering PDD services (Interview 52).

Private consultancies also do not regard the government affiliation of centers as an advantage, because it would not necessarily be useful, since projects must be approved by the central government alone, and because of the inefficiencies mentioned above. Their alleged role as market facilitators has been criticized by many private market actors. One international buyer commented: 'No, they are not facilitators. It's all commercial, they are brokers, they work for a fee' (Interview 27). When asked about the difference between private project developers and the government-led CDM center, the deputy director general of Gansu's Science and Technology Department elaborated his view:

> In the beginning, the CDM in Gansu was launched by the government and we established the CDM center as a trial. When the trial was successful, private companies also became project developers. As a government entity we treat these companies equally and provide them with our services. In absolute terms, the difference is not big. For a government entity the main tasks are CDM information dissemination, initiating and coordinating the CDM; for the private company the main task is CDM project development. (Interview 50)

Appropriate timing of public and private market intervention

The timing of a CDM center's intervention has turned out to be another important factor determining its effectiveness. That timing is of course partly related to the date of the center's establishment. In China, the Ningxia CDM center was the first to be established, in 2004, the Gansu and Hunan centers were initiated in 2006, and the Yunnan center was a latecomer, being established in January 2007. Of the total of 27 centers, many others are still in the planning stage. The case studies on the four provincial CDM centers revealed that the timing of a center's establishment makes a difference to its effectiveness. It is a necessary but not sufficient explanatory factor, because latecomers are also able to catch up in performance with the earlier entrants. For example, while the Yunnan CDM center certainly suffered from its late establishment, at a time when market turf had already occupied by private consultancies in Yunnan, the Hunan CDM center was able to overtake its private competitors despite a late start.

Empirical findings verified the assumptions of innovation theory with regard to the most conducive timing of public and private interventions for market development. Innovation theory generally recommends a 'market-push' by public actors in the early phase and 'market-pull' by private actors in the later phases. The findings from the four case studies confirmed this prediction. Centers that became active in the early phases of market development were usually able to establish themselves as their province's CDM knowledge hub, and also did well in terms of project development. Centers that entered the CDM market in a later phase of its development were instead faced with high competition from private consultancies that had already claimed their market shares. A high impact could be achieved in the first phases of awareness-raising and capacity development, but only a mixed impact in the early phases of product diffusion.

The initial assumption that, once the CDM has diffused within the market and the market has gained sustainability, there should be no further need for state agency, and the expectation of CDM centers to be able to phase out their intervention strategies and withdraw from the market, could not be confirmed. Besides considerable market development in all four provinces, CDM centers continue their activities, although they have shifted their focus towards offering for-profit CDM project development services. A crucial but unresolved issue in the provincial CDM markets is the question of whether, and when, the CDM centers are supposed to withdraw from the market as semi-public actors. From a theoretical point of view, their lingering and interference in the market might cause market distortions, as their advantageous position in terms of public financial support might be thought to crowd out private actors offering the same services. These kinds of development have also been observed for other market-facilitation organizations, such as BDS centers, which relied on government support but delivered private goods and services (Marr, 2004: 8).

Last but not least, empirical findings also confirmed the notion of innovation theory that the diffusion process itself is not necessarily a linear process, with one phase of diffusion following another, but a complex and iterative process characterized by the operation of several feedback loops (Fan et al., 2009: 29; Maleki, 1991; Edquist and Jacobsson, 1988). The provincial CDM diffusion process also showed that linear diffusion occurs only for small segments of the market (for example, for the hydropower sector of one province). For each new sector or region of diffusion, the process has to start from scratch, although there are many feedback loops between these individual diffusion processes. There are feedbacks between sectors (for example, hydropower project owners sharing their experiences with companies interested in energy-efficiency projects), between provinces (for example, CDM center staff from Hunan providing training for their colleagues from the Anhui CDM center), and within different levels of the process (for example, many provincial companies received information and training at both the international and national levels). The capacity and experiences gathered on the CDM market are also transferable to the VER market, or to other diffusion processes for products similar to the CDM.

Implications of findings for innovation theory

The findings from the intervention strategies of the Chinese CDM centers partly confirm the assumptions derived from innovation theory. If one conceives of the CDM centers as public actors, their behavior fits the assumption that first public actors 'pull,' and later private actors 'push' innovation along the diffusion curve (Meyers and Marquis, 1969: 60; Mowery and Rosenberg, 1979: 104). What happens, however, if we understand CDM centers not as mere public actors, but as hybrid actors that adopt different proportions of public and private character over time?

At the level of the empirical research underlying this book, findings concerning the effectiveness of the CDM centers reveal very mixed results. One major research finding is that CDM centers' effectiveness in CDM diffusion is strongly influenced by their organizational structure, which is always a form of hybrid on a continuum, from being a part of the provincial government, via being a semi-public agency with profit-making opportunities, towards being an independent private company. If they are a part of the provincial government, they perform well in the early market phases in terms of information dissemination and capacity development, but are not competitive with private consultancies in the area of project development in the later market phases. If CDM centers consist of both a public and a private entity, they tend to concentrate on profit-making CDM project development services, and carry out only the essential aspects of information dissemination and capacity development. Figure 5.2 shows how CDM centers act according to their hybrid character: their public halves do indeed provide public goods such as information, knowhow and capacity development, while their private halves tend to concentrate increasingly on offering for-profit activities, such as paid consultancy services for CDM project development.

One major empirical finding is thus that the Chinese CDM centers do not resemble the ideal type of diffusion catalyst and market facilitation organization outlined in the textbooks, but – due to their hybrid character – deliver only mixed results in providing the public goods of CDM information and knowhow to nascent CDM markets. Also, contrary to initial assumptions, there were even some cases of private business actors who had already become active in the early market phases in providing information and learning opportunities, because they were interested in establishing early contacts with potential customers for CDM project development. This hybrid character of the CDM centers was not envisaged and cannot be explained by innovation theory, probably because this theory was build on experiences from OECD countries, which tended to show a clearer public–private dichotomy.

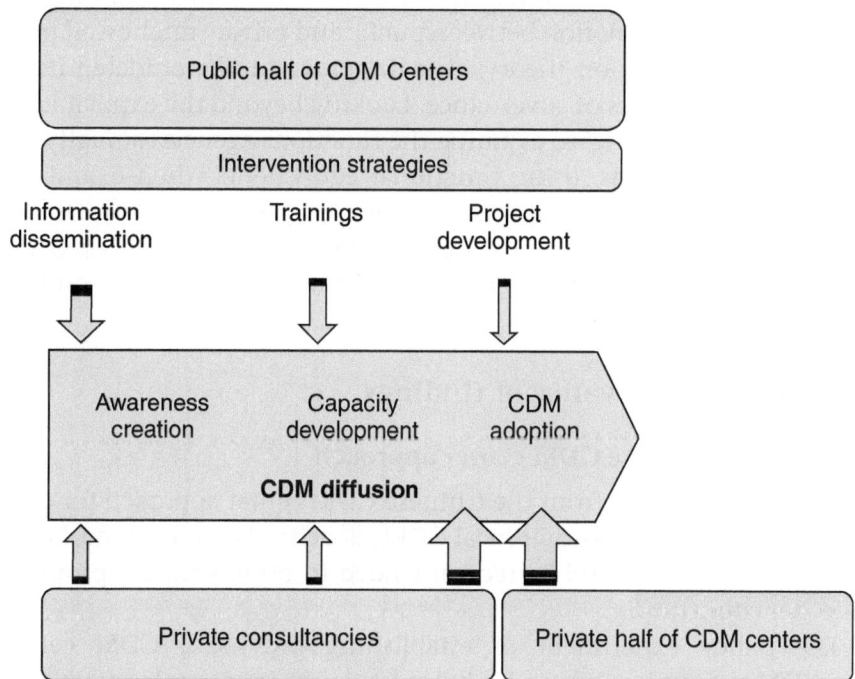

Figure 5.2 Effectiveness of the CDM centers for CDM diffusion and market development

Innovation theory has not yet incorporated the possibilities presented by hybrid actors. This is probably because the core of innovation theory was built in the 1960s and 1970s, based on market experiences in the USA and Europe. At that time, the distinction between public and private actors was much more accurate than it is now, even in non-OECD countries. The emphasis of research was on the comparison of innovation systems of OECD countries which have economies that rely on similar principles and types of industrial organization. This narrow focus has been acknowledged as a limitation of innovation research (Nelson and Rosenberg, 1993: 507; Liu and White, 2001: 1,092). That is why innovation theory's assumption about the role of public and private actors is not valid for non-OECD countries, which are often home to hybrid actors. As my research findings suggest, it is not so much the label of 'public' or 'private' held by an agency that determines its role in innovation diffusion (a point raised in relation to China in Liu and White, 2001: 1,093). Instead, it is its organizational structure, particularly its ownership model, that determines the possibilities for an organization to become effective in its mandate to diffuse innovations. Organizational structure, however, remains a factor that is underdeveloped as a variable to explain the effectiveness of diffusion catalysts. The traditional distinction between public and private might well prove obsolete for innovation theory, as it has proved to be outdated in the debate on new modes of governance. Looking beyond the explicit labels of 'public' and 'private' to examine the functions agencies actually perform (in other words, using functional equivalents when examining actor groups and their impacts, as proposed for example by Draude, 2007) might be one necessary step in adapting innovation theory to the changed circumstances of contexts that have moved beyond the public–private dichotomy.

The Practical relevance of findings

How to improve the CDM center approach

What can be learned from the Chinese CDM center approach for other countries in need for greater local CDM capacity? Is there an ideal CDM center model that could be recommended to other Chinese provinces or other countries?

The policy experiment of establishing provincial CDM centers for CDM promotion has to be judged as only moderately effective in achieving its objectives. Analysis shows that the CDM centers' roles do not fit the roles for market-facilitation organizations envisaged in the literature (see Chapter 3). CDM centers saw profit-orientated PDD development as their core activity, and tended to neglect the provision of public goods such as CDM information and CDM capacity development. CDM centers also limited their target group to project owners and government officials, excluding an important market segment – other project developers. CDM centers thus became almost equivalent to other project developers.

Despite its flaws, the CDM center approach is recommendable for other CDM host countries that need to build up CDM capacity at the local level and kick-start their nascent CDM markets through public intervention. However, the following options might serve as initial ideas on how to improve the way in which CDM centers achieve their goals:

1. establish a clear mandate for CDM centers' business development and an exit strategy once their mission has been accomplished;
2. improve their institutional reliability and accountability;
3. establish a quality standard for their services;
4. agree on a national CDM center selection procedure.

A clear mandate, focusing either on the centers' facilitating role in encouraging other private sector actors to provide goods and services or on their role of offering these services directly, would clarify their organizational development strategy. Resolving the role of the CDM centers is one important task for the future. This book's research has shown that there is a great discrepancy between the assigned role of the centers as market facilitators and their actual role as almost company-like project developers. What kind of role they should adopt is an important decision not only for the Chinese CDM centers, but is usually an essential question for other market-facilitation centers as well. For example, Levitsky notes unease on the part of various stakeholders concerning the assignment of roles within BDS centers. The centers should only take on the role of facilitator, meaning that they would be transitional in nature and phase out their services once the market had reached a certain stage of development. If centers supplied services themselves (not only facilitating the supply of services by other, private companies), they might offer unfair competition to private firms, and eventually distort the market. In order to circumvent such market distortions, BDS centers should focus only on developing a market for private sector consultants and their services, instead of becoming long-term direct service providers themselves (Levitsky, 2000: 9). A similar recommendation would make sense for CDM centers initiated to kick-start local CDM markets, and to withdraw or change their objectives once the market had developed to a certain point.

An exit strategy is needed for phasing out their activities once CDM diffusion and market development have reached a point at which a degree of public support is no longer necessary. Experiences with BDS centers reveal that this is a healthy approach if donor financing drops to around 25 percent of costs after one or two years of operation, while the remaining 75 percent of costs are to be covered directly by income generated from the center's activities (Levitsky, 2000: 6). CDM centers thus face two challenges ahead: the first is to move beyond the phase of donor programs (with financial support); the second is to survive in a market environment, with heavy competition from private consultancies. The first challenge is mostly about financial sustainability. As the matter of financing is closely linked to the autonomy of an organization, the decision on which financial sources to use is also a decision about who (donor, government, or client) an organization should receive incentives and instructions for its business operations from. The second challenge is about finding an adequate market niche and an appropriate strategy. This would entail reaching out to one's own target group, offering services that meet demand, being perceived as legitimate and capable, and building up client and partner networks. In short, the organization needs to learn continuously from its experience and remain responsive to the changing needs of clients and open to innovations in its intervention strategies (UN, 1982: 20). In the Chinese case, such an exit strategy is not available to CDM centers, because it is in the interest neither of the foreign donor to lose a valuable organization for project sourcing nor of the provincial governments and the CDM centers themselves to forfeit the attractive revenue opportunities from the CDM business. If CDM centers wished to continue their work, they could focus instead on new niche markets in which public intervention was still needed to push innovations towards diffusion. Such niche markets might be either untapped CDM sectors or other CDM-like products, such as VER projects on the voluntary market; MDG projects, which should have more sustainable development benefits; or publicly funded emission-reduction projects. Only a handful of CDM centers have yet managed to achieve financial sustainability and become independent of Sino-foreign donor programs. Most of the 27 CDM centers are still in the process of initiating the operations for which they have support from foreign donor programs. There are also provinces without existing or planned CDM centers: Jiangxi, Guangdong, Fujian, Heilongjiang,

Hainan, Tianjin, Chongqing, Shanghai, and Beijing (Interview 10). According to representatives of MOST, Shanghai and Beijing will never have CDM centers because they have no need for them, already having plenty of private CDM consultancies (Interview 3).

Interviews with Chinese market participants have revealed many concerns about the reliability of the CDM centers and their quality of work. For example, it is not easy for outsiders to understand which institution or company is the official CDM center of a province, and which one merely claims to be (Interviews 10 and 27). This situation is further complicated because some provinces, like Sichuan and Xinjiang, even have two official CDM centers, because of internal decision-making problems (Interview 14). In addition, it is confusing if one person hands out two business cards, claiming to be the director of both the provincial CDM center and a private consultancy. While this can indeed be the case for some provinces – Ningxia or Hunan, for example – foreign companies are especially confused about this practice, which, from their point of view, undermines the credibility of their Chinese business partner (Interview 27). Having transparent standards and an open nomination system for provincial CDM centers may therefore build greater trust.

Another question is that of how to ensure a common quality standard for the services offered by provincial CDM centers. Several interviewees noted the varying quality of PDDs developed by the CDM centers. For foreign buyers, in particular, it would be difficult to keep an overview of which CDM centers they could rely on and which ones offered insufficient PDD quality. Interviewees thus recommended having reliable and comparable quality standards for the PDDs developed by the CDM centers (Interview 10). One possible means of achieving such comparable quality in services could be to have the centers become provincial franchises of a supervising CDM center. This would eradicate the competition among the centers, and would prevent the low-quality services of the 'black sheep' spoiling the reputation of the good performers. If 'provincial CDM center' were to become a label for these semi-public agencies, guaranteeing their reliability as a business partner, it would be necessary to ensure that the various centers also subscribed to a common quality standard. So far, even an analysis of the four provincial CDM centers has revealed that there are large differences regarding the type, scope, and quality of services offered. A barrier to this one-size-fits-all approach is of course the heterogeneity of the Chinese provinces and the degree of autonomy that has to be granted to each provincial government in setting up such centers.

There are no clear guidelines from central government on the selection procedure for provincial CDM centers. On the one hand, some government representatives argue that it should be left to the market to 'pick the winners' among CDM centers (Interviews 2 and 10). On the other hand, there seems to be some tightening of government control over the proliferation of self-proclaimed provincial CDM centers that often turn out simply to be private consulting companies without a connection to the government (Interview 27). One buyer, however, also stated that he had experienced a situation in which two entities that declared themselves to be the provincial CDM centers were in a mutual standoff, each denying the other's claim to be the official CDM center. It subsequently transpired that both had some sort of connection to the Science and Technology Department of their province (Interview 27). This example might also show that provincial governments do not always restrict themselves to one CDM center. Competition between different provincial departments to take the lead in relation to the CDM might end in similar compromise solutions, resulting in the coexistence of two centers. Related to this point is the question of how the selection process for CDM centers in the provinces might be improved. Until now, most CDM centers have been selected in a top-down manner, although sometimes two centers are picked – for example, in Sichuan – because the provincial government does not speak with one voice, but, rather, different departments have their different favorites. Guizhou province, however, chose a bottom-up approach in selecting a CDM center: it was selected through a tender procedure initiated by the donor (the UK government) involving several competing institutions (Interview 11). An interesting task for future research would be a comparison of the top-down approach with a bottom-up approach in which institutions can compete for the status of a CDM center, which is then awarded government support and donor financing.

In short, discussion of the strengths and weaknesses of the current Chinese CDM center model reveals that the approach of setting up centers for local CDM market facilitation can be recommended for other countries, but that there is still much room for it to be improved.

Future business strategies for CDM centers in China

The future outlook of CDM centers in China depends on making decisions concerning their organizational orientation, overcoming their present conflicts of interest, and diversifying their services and finances. One possible means of moving away from their current focus on CDM development and towards providing the public goods of information and capacity development to the market, while also maintaining profitability, would be to shift their activities to the CDM niche markets. This is also a popular strategy of donor agencies from Annex I countries, which are not supposed to spend ODA directly on CDM projects but nevertheless wish to support their national industries in CER purchasing. They therefore focus on indirect CDM support measures that are not yet commercially attractive for private business actors to undertake. Redirecting the CDM center's activities towards such niche markets would of course also mean steering a difficult course between working for public goods and maintaining financial viability. There are some projections of what such engagement in niche areas of the CDM market in China might look like. For example, the director of the Gansu CDM center proposed that the centers should concentrate more on capacity development measures in sectors whose CDM potential has not yet been effectively tapped. Instead of establishing CDM centers everywhere, the centers should be concentrated on selected sectors and topics – for example, on energy efficiency (Interview 49). Although the representative of the Gansu center was referring to sectors within provinces, the idea of establishing 'centers' which do not have regional expertise, but have already acquired sectoral expertise: for example, there are 'CDM centers for the coal industry,' subsidiaries of the China National Coal Association. Diversifying services might also entail extending services to other provinces (which is already done, for example, by the Hunan CDM center) or to other countries (a strategy planned by the Ningxia CDM center, which wanted to offer consulting services in African countries). As a first step in this direction, an initial CDM training course for African participants took place on November 10, 2007 (Interview 3).

What are the options for diversifying services on the local CDM market in China? What are the current and likely future demands for CDM center services? CDM capacity has improved considerably in China. This means that, for the business-as-usual CDM projects – projects with clearly established methodologies in well-known sectors with many existing instances – project development is no longer a challenge. On the contrary, PDD development is becoming a standard manufacturing procedure, with rumors of researchers recruiting students to do 'copy-and-paste' PDDs as term assignments. Instead, the development of methodologies for new project types and sectors remains a bottleneck, and the validation period is being prolonged due to overworked DOEs and a busy DNA. The quality of PDDs is insufficient, as they either include unsuitable arguments or mismatched data, parameters, and monitoring plans (Heggelund and Ying, 2008: 18).

Most of the interviewees had already heard about the possibility of running VER projects, but have no concrete knowledge about eligibility criteria, the possible pitfalls, or potential cooperation partners. Special trainings on VER requirements would help to fill this gap.

A similar demand exists for better information on programs of activities (PoAs). This would include training, but also further research about this new project type's feasibility is needed. Several interviewees also expressed their interest in learning more on the financial aspects of the carbon market, and were especially keen to improve their ERPAs negotiation skills and gain a better understanding of the requirements for the second and third crediting periods. Interviews revealed that many project developers were becoming increasingly interested in the secondary carbon market, probably in an anxiety to receive their slice of the global carbon finance 'cake.'

Even by beginning of 2008, local banks still had little knowledge about the CDM, and often did not take CER revenues into consideration for the purposes of loan approval. This lack of interest on the part of banks applied not only to CDM projects, but to environmental investments in general (Spofford et al., 1996). Until now, it has often been the project owners who had to explain the CDM to banks, despite their own limited knowledge. A tailor-made approach for financial institutions in CDM capacity development is thus overdue.

The central government has a clear point of view about the future tasks of CDM centers: their long-term strategy should be to extend their work to other areas of climate change. Centers are supposed to diversify their activities. In addition to doing CDM promotion and project development, they should also focus increasingly on adaptation needs and measures. The centers should also investigate their province's potential for greenhouse gas reductions and advise their provincial government on policies to promote mitigation projects – especially CDM projects that deliver special sustainable development benefits (projects that fit into China's three priority areas of renewable energies, energy efficiency and methane avoidance) (Interviews 2 and 3). Such a diversified intervention strategy is already suggested for centers participating in the UNDP's MDG program, which supports the establishment of CDM centers in the western provinces (Interview 1).

Representatives of the central government also voiced a surprisingly market-orientated approach to how they thought CDM centers could achieve sustainability: 'It's all up to the market to decide; and

CDM centers have to fight for their survival in the market by themselves' (Interview 1). Accordingly, the central government would only support the CDM centers in the initial phase of their establishment. Future funding should therefore only go to CDM centers that are newly established and still need assistance to strengthen their capacity (Interview 1). The central government, in cooperation with foreign donors, thus provides the seed funding and initial capacity development for centers' staff, but how they make use of that assistance is up to them: 'Not all 27 centers will be successful and continue to exist. We support them if they choose CDM to be their core business area' (Interview 1). The central government regards the ability to compete on the CDM market as one core ingredient for success. This ability would depend to a great extent on the importance CDM centers place on their CDM work in relation to other work – work that is not climate-related but has been requested by their provincial government. The lack of qualified staff, especially those with language skills, represents a problem for several CDM centers (Interview 1). The competitive advantage of the CDM centers in contrast to private consultancies consists in their in-depth knowledge of their provincial situation and their close contact with the government. There is not much competition between provincial CDM centers, because they primarily regard private consultancies as their competitors. Instead, many CDM centers cooperate horizontally for CDM project development, and for shared market entry into a new province, and they seek advice from each other. For example, the Hunan CDM center advised the Anhui CDM center on its operational structure, and gave it advice on project development (Interview 1). The final vision for the CDM centers is for them to become operationally independent entities that focus their work on CDM capacity development (Interview 1).

The future plans of CDM center directors contrast sharply with such visions. When asked about their future strategy for business development, representatives of the CDM centers stated that they mainly wanted to expand their PDD services – for example, either by entering new sectors in their own provinces or by sourcing projects in other provinces. An aspiration was expressed to go beyond PDD development and enter the international carbon market as CER traders, or even as buyers of their own projects, by establishing joint ventures with foreign companies. Most of them regarded the time for CDM capacity development in their province as coming to a close, and preferred instead to focus on the CDM business. Some of the CDM centers were planning to continue offering CDM training for externals: the Hunan CDM center has a cooperation agreement to train staff of the Anhui CDM center; and the director of the Ningxia CDM center is even considering offering his training and PDD development services in Russia and in African countries. In line with the centers' vision is the opinion of a representative of the China Association for Renewable Energies, an industry association, which demanded that the CDM centers should be fully commercial and that there should be no political or bureaucratic interference in their work (Interview 19).

While the positions of the central government and the CDM center representatives differ on the question of the core tasks of the centers, they share the impression that the ability to survive in a market with heavy competition is the main challenge to the centers' sustainability. However, if CDM centers become more independent in time because they rely on their own profits and less on government support, chances are high that they will also continue to focus on CDM project development and not become more active in activities with a greater public-interest dimension, such as providing advice to the provincial government on mitigation and adaptation issues.

Observing the establishment of 27 subnational CDM centers in China, the thought arises whether they are the first indicators of a possible decentralization of CDM approval procedures in China. Although China has gone through several phases of decentralization and re-centralization since the political reforms of the Deng area, it is too early to identify the engagement of local government entities in the CDM as a form of decentralization in the strict sense of a 'shift in decision-making and spending power from central to region and local governments' (Campbell and Fuhr, 2004: 11). There are of course advantages to decentralization, such as achieving a better match of public services to local demands and preferences, and 'build[ing] more responsive and accountable government from below' (Fuhr, 1997: 2); but the coordination of the individual CDM policies of 31 provinces and regions would also require great effort. Eventually, these centers might grant CDM approval themselves, based on their superior knowledge of the local situation. But this possibility of CDM centers being the pioneers of the decentralization of the CDM process was disputed by the representatives of the NDRC and of MOST. The authority for CDM project approval would rest solely with the Chinese DNA in Beijing (Interview 1). Although government officials deny the possibility of decentralizing this mandate to the provincial level in future, this possibility has been taken into consideration by representatives of the provincial CDM centers (Interview 19 and 71). In a partial response to those aspirations, a representative of MOST even expressed his concern that some of the provincial CDM centers might become overambitious in their wish to take over project-approval functions from the DNA:

> A centralized means of project approval is quicker and less complicated. Otherwise the provincial governments would play a more prominent role, which would make everything more complicated and time-consuming. Some provincial CDM centers tend to work closely with their provincial governments, but we discourage them from doing so. This causes 'difficulties.' (Interview 1)

Transferability of the CDM center approach to other CDM host countries

In his well-received report on the 'economics of climate change', Stern called for an international effort for countries to learn each other's best practices in policy design for low-carbon economies (Stern, 2006: 4). The obvious question is what, if anything, can we learn from the Chinese experiences with CDM centers. Although the Chinese situation is unique, there are policy features of its CDM centers that are worth testing in different contexts, and there are many candidate countries for such an endeavor. On the one hand, countries like India and Brazil, which have comparatively large CDM markets, still need to strengthen their CDM structures and capacities on the ground in order to continue with large-scale CDM project development. Although CDM diffusion in these countries is already well advanced, there are still geographically remote regions and new sectors in which CDM awareness and capacity are still insufficiently developed. On the other hand, there are countries that have high CDM potential, but still lack CDM awareness and capacity at the national level (for example, most African countries and least developed countries). In both country categories, there should be a high demand for market-facilitation organizations such as CDM centers.

In principle, innovative policies can spread 'horizontally' to other equivalent levels, 'vertically' to lower or higher levels, or out to other countries (Campbell and Fuhr, 2004: 62). The CDM center approach has already spread 'horizontally' among the 27 Chinese provinces, so this section focuses on whether this approach could also be transferred 'horizontally' to subnational levels in other countries. The process of transferring practices that have proved successful in one context to another is incremental, and should at least include five basic steps: first, defining the needs of the 'recipient country'; second, checking for possible practices from other contexts; third, analyzing the latter for their context particularities; fourth, adapting them to the contextual requirements of the recipient country; and fifth, monitoring and adjusting their performance over time. This process of adapting good practices for one's own purposes can be summed up in the motto, 'Scan globally, reinvent locally' (Fukuda-Parr et al., 2002: 18). Besides the process itself, a successful transfer of good practices is also very much dependent on the initiator and agent of such a process (Campbell and Fuhr, 2004: 62). Because the agent's capacity and financial means, but also his or her ownership of and commitment to the process, are important, one promising approach for transferring good institutional practices is the 'twinning' of two institutions – one from the recipient country, one from the country from which the practice is going to be transferred. This twinning approach has the advantage of combining training with technical assistance, with the prospect of long-term cooperation between the two participating institutions (Ouchi, 2001: 7). Also important in such a process would be the ability of the agents to dissolve the many points of resistance to changing the established practices of the existing institutions in the recipient country (Minogue, 2001: 35). Applying such a twinning approach, CDM center–like institutions could be set up in other countries with the close involvement of Chinese CDM center staff.

What can be learned from the Chinese experiences for other countries that have local capacity-development needs for their full participation in the CDM? One major difficulty of transferring good practices from China is in accommodating the particularities of the Chinese context. Without getting into the controversial debate how much the Chinese context is unique and incomparable to those of other countries and cultures, a few features of the CDM center approach that are indeed due to its Chinese origin should be highlighted.

Probably the most striking feature of Chinese society is the distinctive state-centrism of its institutions – a feature that is hard to find in other countries interested in replicating Chinese experiences (Oi, 1995: 1,132f). Three crucial 'Chinese characteristics' can be identified in the Chinese administrative system (Minogue, 2001: 35f): first, the importance of loyalty to the Communist Party and the continuance of a political cadre system that is in tension with a more professionalized performance-based approach (Tong et al., 1999; Zhou, 1995); second, the strong attachment to personal networks and reciprocal obligations; and third, the country-wide imperatives of modernization and development that shape individual values and behavior (Zhou, 1995: 448). This particular Chinese setting was also confirmed by interviewees in research for this book, who remarked that, compared to other countries, the main distinctive aspect of the Chinese CDM market was its strong government and its ability to enforce policies (Interview 49).

A second particularity of the Chinese CDM center approach is its intangible *shiye danwei* basis, which I have already discussed at length. Although such semi-public, semi-private agencies exist in other countries as well, they may operate very differently. On this issue, it might be helpful not to try to find possible matches to the Chinese organizational form, but instead to look for functional equivalents (for a good overview of the applicability of this approach, see Draude, 2007), or institutions that might look different in their structure but which share a strong government attachment. A government attachment, however, would only be useful if the public administration of the recipient country shares the feature with the Chinese administration of being able to support the facilitating center in communicating CDM information to potential project owners. If the public administrations are not able or willing to do so, their involvement might turn out to be an obstacle to the CDM-facilitation process.

Referring not directly to the CDM center approach, but to China's generally positive CDM development experience, the NDRC is proposing to enable other countries to draw upon China's success story, and recommends a focus 'on countries in Africa but also other Asian, Central Asian and Middle East countries that may be in their respective upstart phases of CDM application' (World Bank, 2010). As a first step in launching this exchange of experiences on the CDM, the World Bank's Carbon Assist Program has organized a 'Clean Development Mechanism (CDM) South–South Cooperation between China and other Developing Countries.' It was kick-started with a seven-day workshop in Hubei Province, China, at which policy-makers, project developers and service providers from countries in Africa, Asia, Central Asia, and the Middle East participated in order to learn from China's relevant CDM experiences (World Bank, 2010).

There are already several approaches in other CDM host countries that are comparable to the Chinese CDM center approach. As a response to the widespread lack of CDM capacities, an array of capacity development programs targeting the sub-national and local levels has emerged. These programs primarily target regions and countries that so far remain underrepresented in the global dissemination of CDM projects (UNEP Risoe Centre, 2008). For example, there are plans to set up a regional CDM center for Africa. The aim of such a regional CDM center which is supposed to be set up in Kenya is 'to create a pipeline of CDM projects in Kenya and in the region' with a vision of becoming

> the preferred regional one-stop shop for carbon buyers and carbon offset investors. The organizational setting of such a regional CDM center would be a public-private mix. On the one hand, associative partners would be Kenyan ministries and development agencies; one the other hand, also privates sector partners like industry and business associations and private financial institutions would be involved. (Njoroge, 2007)

Another program that is comparable to the Chinese CDM center approach is the idea of setting up CDM nodal agencies in India, which are supposed to be state-level sub-branches of government departments. Like the CDM centers, the nodal agencies are intended to promote the CDM at the local level. These agencies have also received support from foreign donor agencies: the UNDP has collaborated with the Indian Ministry of Environment and Forests in a capacity-development program to support pilot agencies in five states: the Punjab Energy Development Agency (PEDA), the Maharashtra Energy Development Agency (MEDA), the Rajasthan Energy Development Agency (REDA), the Environment Protection, Training and Research Institute (EPTRI), and the Environment Management and Policy Research Institute (EMPRI) (TERI, 2004: 25). Much like those of the Chinese CDM centers, the activities of the Indian nodal agencies will include the identification of CDM projects; facilitation of project-development services; provision of capacity-development measures to CDM stakeholders; dissemination of CDM information; analysis of local sustainable development priorities; maintenance of a database of technologies, consultants, operational entities, funding agencies, CER buyers, and so on; and updating of stakeholders with ongoing changes in the CDM modalities. It remains to be seen whether the Indian nodal agencies will indeed take up the role of facilitating their state-level CDM markets, or will instead focus primarily on project-development activities. Taking into consideration their organizational status, which seems to be that of either a government department or a research institution, the assumption that they may be less efficient in project development but more focused on providing facilitating services than their Chinese counterparts gains greater weight.

Part III

Implications of Hybrid Actors for Environmental Governance

6
The Blurring of the
Public/Private Boundary

CDM centers as hybrid actors of mixed effectiveness

CDM centers can be seen as policy experiments for determining how much state intervention is necessary and conducive to the functioning of market mechanisms in China. Innovation theory and practical experience with other kinds of market-facilitation organizations suggest that state interventions can have positive effects on market development if the market-facilitation organizations adopt the role of catalyst for the diffusion of the CDM, for example by providing several public goods, such as information and knowledge for private market actors. Empirical findings from the comparative case studies reveal that the CDM centers were effective only in the provision of such public goods in the early market phases, but later on opted to concentrate on the pursuit of private goods (mainly service fees for project development). These results were disappointing in two respects. First, the Chinese case studies showed that the CDM centers did not resemble the textbook ideal of the market-facilitation organization, but rather actors that – after the initial provision of essential public goods – pursued profit and neglected their public mandates. Second, CDM centers functioning as hybrid actors also disappoint the hope that hybrid actors might offer a new solution to governance problems, supplanting inefficient public administrations in the provision of public goods.

The purpose of this chapter is to discuss the findings of the case studies on the Chinese CDM centers in a broader perspective. Some of the main phenomena observed, such as the variation in effectiveness of the four CDM centers due to the timing and scope of their interventions, can be explained by innovation theory; but there are also unexpected observations that it cannot explain. This chapter aims to reach an understanding of the phenomenon of the hybrid nature of the CDM centers, and their questionable performance in the provision of public goods. Thus, it reviews different theoretical approaches for explanatory factors in order to answer two questions: Why do hybrid actors occur? And what do they deliver? It is to be hoped that examining the Chinese hybrid actors will provide some insights about functional equivalents to purely public or purely private actors. Although a note of caution must be sounded on the methodological possibility of generalizing the findings from four case studies in the very specialized field of provincial CDM markets, these policy experiments are worth discussing for how they fit into the existing debate on the transition of environmental governance in China.

For an answer to the first question – why do hybrids emerge? – I review the debate on new forms of governance and the literature on transitional economies. Hybrids are a new topic in the governance debate, but are familiar in the literature on transitional economies. The two approaches offer different explanations: the first approach explains hybrid actors as a consequence of the blurring of the tasks of private and public actors in governance in OECD countries; the second approach explains hybrid actors as a consequence of the blurring of ownership rights in transitional countries. Merging the insights of these two disciplines on the role of hybrid actors may help to develop concepts for analyzing hybrid actors further, by taking into account their different systemic contexts as intervening variables.

On the second question – what do hybrids deliver? – literature review on the achievements of hybrid actors in China reveals that the performance of the CDM centers is in line with a general tendency for hybrid actors: they tend to neglect their mandate for the provision of public goods, while concentrating their efforts on the generation of private goods. Their mixed performance is explained by their conflicting mandates, their inherent conflict of interests, and the lack of control exercised by the public stakeholders whose responsibility is to monitor and steer their performance.

Finally, it is argued that the blurring of the private–public dichotomy only mirrors the blurring of the market–state dichotomy. The argument is made that the new roles adopted by local bureaucracies as market actors (i.e. CDM centers) are just one more example of how the role of the state is changing in China – from being a regulator relying on top-down control, the state has become a facilitator of the market economy, and eventually taken on the role of entrepreneur itself.

The emergence of hybrid actors

In Chapter 5 a range of advantages and disadvantages of having either a public or private status as an agency for CDM market facilitation was discussed. The question why CDM centers take a hybrid form has not yet been explained. Also unclear is the definition of what constitutes a 'hybrid actor.'

Hybrids are new in the governance debate...

In the debate on new forms of governance, it has become fashionable to speak of hybrid actors, hybrid forms of governance, and hybrid regimes (Erdmann, 2002; Rüb, 2002; Lemos and Agrawal, 2006; Erdmann, 2007; Bünte, 2006).

Hybrid actors are considered a new phenomenon in a debate that used to rely on the binary categorization of types of actor – state- and non-state actors, public and private actors, profit- and non-profit actors, and so on. These distinctions made sense for inferring different causal mechanisms from these different kinds of actor constellations on governance outcomes – for example, 'governance by government,' 'governance with government' and 'governance without government' (Zürn, 1998; Rosenau, 1992; Rhodes, 1996: 652). State actors have been traditionally seen as the guarantors of governance outputs, but their alleged inability to deliver effective and efficient governance provoked many accounts of 'government failure' (Jänicke, 1990). With the simultaneous emergence of private actors – both for-profit and non-profit actors – in governance constellations (Cutler et al., 1999; Hall and Biersteker, 2002; Cashore et al., 2004; Keck and Sikkink, 1998), the state soon came to be seen as being on the retreat (Zürn, 1998). Instead, governance styles have emerged in which the boundaries between and within public and private sectors have become blurred (Stoker, 1998; Haufler, 1995); this is also true of the boundaries between for-profit and non-profit private actors (Wolf, 2008: 234). Lately, research on new forms of governance has been able to observe the emergence of 'hybrid actors' in various political fields (Schuppert, 2008). Consequently, many authors are calling for the abandonment of the outlived 'private–public' dichotomy (Rhodes, 1996; Pierre and Peters, 2000), especially when analyzing governance actors in areas outside the OECD (Draude, 2007: 7).

For a few years now, this general debate on new forms of governance, the emergence of hybrids, and the old-fashioned dichotomy between public and private actors has been adopted and further elaborated by researchers working on climate governance (Pattberg and Stripple, 2008; Jagers and Stripple, 2003; Andonova et al., 2007). Climate change as a transnational issue is prone to multilevel, multi-actor forms of regulation, and the CDM is a prototype demonstrating that all phases of the policy cycle can involve public and private actors, and mixes of the two (Haites and Yamin, 2000; Corell and Betsill, 2001; Streck, 2007). Due to the lack of adequate state capacity in many countries, hybrid modes of governance such as public–private partnerships and co-management of CDM projects are often seen as promising options to tackle global climate change (Lemos and Agrawal, 2006: 315). Hybrid actors already play a crucial role in the implementation of the Kyoto Protocol: for example, certification companies for CDM projects are sometimes private companies with a strong attachment to their national governments, and governments themselves act as market players when they buy CERs through their public purchase programs.

Despite the new publicity about hybrids, a definition of what constitutes a 'hybrid actor' is still outstanding, although there are several suggestions of how to define 'hybrid regimes' – for example, as 'political regimes which have the features of two regime types but in which none of these features is dominant' (Schubert and Tetzlaff, 1998: 20; my translation). Critical in this definition is the interpretation of how much of one feature must be present for one to say it starts to 'dominate' another. At exactly what degree of, for example, operation as a private company does the actor stop being a 'hybrid' and start being a 'private' actor? The imprecision of such distinctions has not yet resolved, and has already been identified by several authors as a research desideratum (for example, see Croissant, 2002: 17).

Theoretically informed tools for analyzing hybrid actors are also lacking, although there are several proposals on how to conceptualize hybrid regimes (Croissant, 2002: 17ff). On the question of forms of 'hybrid governance,' Draude has proposed the use of the heuristic device of looking for 'functional equivalents' to our Western understanding of governance when conducting research in areas of limited statehood (Draude, 2007); and Schuppert has suggested differentiating between three forms of hybridization of governance – namely according to their legal, organizational/institutional, and functional dimensions (Schuppert, 2008: 34). But the empirical findings of this book have already shown a causal relationship between the three dimensions suggested by Schuppert: the legal basis of the CDM centers, with respect to public versus private ownership, and their organizational form – whether they operate like public or private entities – seem partly to explain their function in delivering public and/or private goods.

What do hybrid actors actually look like? Although research on how to conceptualize hybrid actors has only recently emerged in the academic debate, there are many examples of hybrid actors throughout social history. They are more familiar, for example, in the form of Chinese merchants, who also contributed to food security; or the East India Company, which represented the British crown and took up governance functions; or the private tax collector – historically in the Roman Empire, and presently in many developing countries (Schuppert, 2008: 18, Badian, 1997; Stella, 1992; Leutner, 2007). Hybrid actors are also well-known today in OECD countries. These take the form of 'agencies' and 'public enterprises' that are non-profit but performance-based entities providing public services for or supplementing the state (Pollitt et al., 2004; Döhler and Jann, 2007; Edeling et al., 2001). On the one hand, they usually receive their mandate from the government, and are accountable for outputs and outcomes; on the other, they operate in a business manner with appointed executives and output-based contracts. For example, New Zealand recently took this outsourcing of public services to private entities to an extreme: it broke up its ministries into company-like business units (World Bank, 2007: 10).

There are several explanations for the emergence of hybrid actors in the OECD. Advocates of the problem-focused, functional approach argue that the complexity and interdependency of today's problems requires a multi-level, multi-actor solution, because no single actor would have the capacity to tackle such problems (on environmental problems, see Lemos and Agrawal, 2006: 311).

A more widespread explanation for this development is the spread of neoliberal approaches designed to save on public spending, to streamline bureaucratic processes, and to infuse entrepreneurial spirit in the government's bureaucracies (Osborne and Gaebler, 1992; Eisinger, 1998), or simply to outsource public service delivery to private actors (Savas, 1979). In short, the diffusion of the 'new public management' paradigm is one explanation for the occurrence of hybrid actors in governance arrangements (Osborne and McLaughlin, 2002). These 'mixed modes or new interplays' of public goods provision are even seen as an attribute of modern statehood (Schuppert 2008: 34). Hybrid governance arrangements are thus the end-product of a process in which formerly independent public and private actors have become increasingly intertwined through the marketization of governance modes (see Figure 6.1).

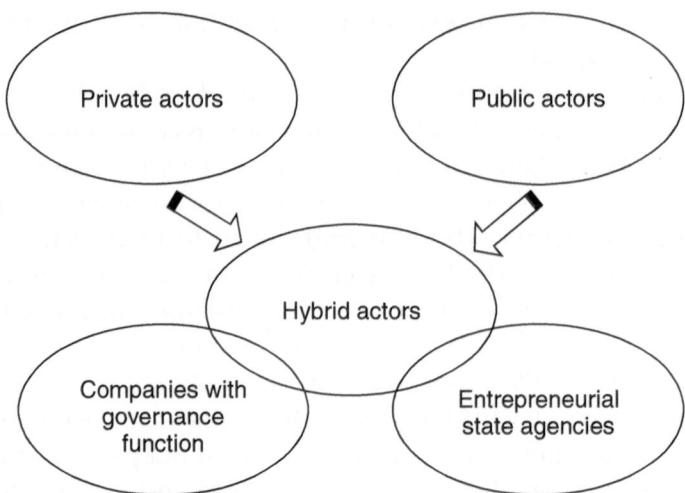

Figure 6.1   The merging of private and public actors into hybrids in OECD countries

...but are an old phenomenon for transitional economies

Although research on hybrid governance actors is not yet well-developed, hybrid actors are a familiar phenomenon for transitional economies. Their appearance and their intended functions are similar to those of their OECD counterparts, but the reasons given for their emergence are very different. While it has become a new paradigm in OECD countries to overcome the public–private dichotomy by turning state actors into enterprises and by making companies responsible for governance functions, the hybrid actors in countries with transitional economies are generally a legacy of their communist past (Nunberg, 1998). Hybrids have emerged in these countries because the process of withdrawal of the state from being intertwined with the economy has taken place only slowly and gradually. Not even former communist countries, following the sudden and wholesale rejection of the communist system, could overcome state dominance at once, because the process itself needed public supervision. A strong shadow of hierarchy usually remains in all public and private spheres (Garcelon, 1997). This excessive presence of the state is a relic of the former communist times, in which the public sector was so expansive that it was almost impossible to distinguish between the state and society, or between public and private actors (Walder, 1995). At least for the Eastern European countries, the beginning of the transition period coincided with a moment when neoliberal concepts of public administration dominated the discourse on the role of the state (Verheijen, 2003: 490). Embedded in this historical context, the greatest challenge for countries with transitional economies is not the merging of public and private actors, as in OECD countries, but their disentanglement (see Figure 6.2). Interestingly, both trends – merging and disentanglement – are advocated in the name of efficiency gains.

While the classical public–private dichotomy no longer holds for the emerging trend of hybridization of actors and governance modes in OECD countries, it does not yet adequately capture actor categories in transitional economies.

Hybrid actors in China

The following paragraphs take China as an example of a country with an economy in transition, allowing a direct discussion of the empirical findings from China in light of the general debate on the performance of hybrid actors in transitional economies. The emergence of hybrid actors in China is embedded in the general transition process from an economic system in which decisions and targets were dictated by the central party state towards one in which coordination is based on market mechanisms and political decisions made at all levels of the state hierarchy (Lee and Lo 2001: 237).

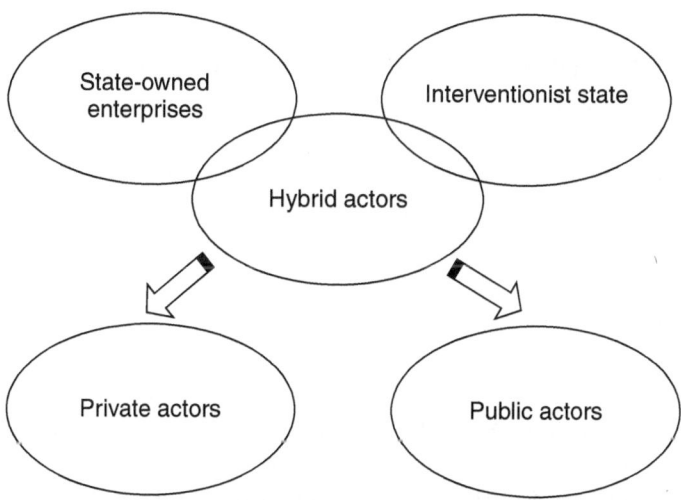

Figure 6.2 Disentanglement of hybrids into private and public actors in transitional economies

CDM centers are just one example of the array of hybrid actors that can be found in many parts of the Chinese economy. Almost all administrative entities in China today have their own hybrid organizations (Lam and Perry, 2001: 28), which they can use both for the provision of public goods and for profit-generation. Because only the hybrid agencies can legally become active in profit-earning activities, many of these organizations have become the 'little treasuries' (*xiao jinku*) of their administrative superiors (Zhang, 1996). For example, many public research institutions and universities have private spin-off institutions that closely resemble CDM centers in their ownership structure and their mode of operation. Usually they conduct for-profit research by making use of the staff and resources of their public mother-institutions, and are often charged with neglecting unpaid basic research (Liu and White, 2001: 1,108). The well-known Chinese Academy of Science operates many of these hybrid spin-off institutions at the local and even at the headquarters level (Hu and Jefferson, 2008: 309). The role of public and private actors is also in a state of flux within the field of environmental governance. For example, public Environmental Protection Bureaus often have private spin-off organizations that which carry out for-profit environmental impact assessments and provide environmental planning services for both the government and private customers (Lo and Tang, 2006: 200). These private customers might include industrial enterprises against which Environmental Protection Bureaus also charge environmental pollution levies, in their function as a public actor – a situation which can of course easily lead to conflicts of interest and corrupt practices (Kwong, 1997).

Similarly analogous to CDM centers are 'state commercial agricultural agencies.' Their labeling as 'state commercial' agencies itself indicates their hybrid character. Research on their performance reveals striking similarities to the observations on CDM centers: they are state agencies that participate in markets like commercial entities. Sometimes they even dominate markets, but 'their ability to exercise market power is often limited by intra-agency competition, new entrants and indirect competition for other markets (Sicular, 1996: 82). Consequently, Terry Sicular concludes,

> in view of changes in the nature and role of state commercial actors, the distinction between 'state' and 'non-state' actors is no longer relevant ... a more useful distinction is that between state agents who are officially charged with policy-related commercial activities, referred to as 'designated' commercial agents, and other, non-designated commercial agents. (Sicular 1996: 82)

This suggestion of distinguishing between organizations according to the public or private nature of their assigned tasks points in the same direction as the suggestion of looking for functional equivalents when analyzing governance forms in non-OECD countries (Draude, 2007), but has also already been advanced for distinguishing between different national systems of innovation (Kogut, 1993: 7).

In the literature on transitional economies, a definition of 'hybrid' has also still to be clarified, but there are Chinese categories available for distinguishing between agencies according to their ownership structure and the locus of their decision-making. The official administrative categories reflect a broad spectrum regarding ownership structure and mode of operation, from purely private to purely public entities. In relation to 'people's enterprises,' the following differentiation is made (Solinger, 1992: 133):

- *minyouminban* – supposedly owned and managed by 'people'
- *guanyou minban* – owned by the state, but managed by 'people'
- *guanmin gongyou* – those jointly operated and owned by the state and the 'people'

Shiye danwei: public service agencies in China

The most common category of hybrid organization in China is that of so-called *shiye danwei* (rendered in the English literature as 'public service units,' 'service organizations' or 'marketized firms'; for a general description of their functions and emergence, see Chapter 1), which may again comprise an array of organizational types that differ in their individual structure. They can be defined as ' "quasi-state" organizations that perform administrative functions and provide public services on an increasingly self-financing basis' (Lo et al., 2001a: 40). Although the history of public service units can be traced back to China's communist times, their importance grew only with reform politics. Public service units were fostered because they were seen as an appropriate organizational form for providing public services while streamlining bureaucracy, cutting fiscal spending, and thereby pushing back the boundaries of the state by trying to do 'more with less' (Lee and Lo, 2001: 6, Lo et al., 2001a). One consequence of the effort to cut public spending was to allow these organizations to generate revenues on their own by engaging in for-profit activities (Lee and Lo 2001: 6). Officially, they are intended to be on an equal playing field with other private companies – for example, when it comes to bidding for public contracts. In the long term, these kinds of hybrid organization are supposed to be transformed into fully fledged independent enterprises (Lo and Tang 2006: 200). In a way, their structure and intended function thus equips them for a role comparable to that of public agencies and public enterprises in OECD countries, some of which combine private ownership with public funding (Burgi, 1999; Schuppert, 2008: 838). In reality, however, public service units in China are heterogeneous, and many hybrid forms are emerging that only partially cater to their original mandate of providing public goods. Nevertheless, with one-quarter to one-third of all public budgetary expenditure going to public service units, they 'constitute a big shadow of the Chinese state' (Lam and Perry, 2001: 20). The growing importance and variety of hybrid actors are typical features of transitional economies. With regard to China, some authors even claim that '[t]his hybridization has become a distinct feature of China's market-oriented reforms' (Tenev et al., 2002: 15), calling it 'institutional amphibiousness' (Ding, 1994: 293), and seeing an 'institutional fluidity, ambiguity and messiness that operate at all levels in China and that [are] most pronounced at the local level' (Saich, 2000: 141).

The most common explanation for the hybridization of actors in countries with transitional economies is the combination of various forms of ownership rights in the process of the privatization of previously state-owned enterprises (Stark, 1996). The hybridization of market and state actors can be explained as being the best solution to the problem of uncertainty in a transitional economy characterized by weak market structures and poorly specified market institutions, such as property rights. In established market economies, business actors' autonomy is clearly defined and based on enforceable property rights. If market participants in transitional economies cannot rely on fixed property rights and established contract-based mediation mechanisms, it makes sense – even from a transaction-cost perspective – deploy other mechanisms as a 'safety net' in business relations. If the state withdraws from its dominant position in the system, networks of personal relations and mutual dependencies are one factor that can reduce the costs of market transactions (Pfeffer, 1972; Pfeffer and Nowak, 1976; Williamson, 1991: 271).

By serving two masters, hybrid organizations thus have some advantages over both purely public and purely private entities, which are especially relevant in situations of uncertainty: they can use the resources and the governance structures of both parties, and they can rely on a public safety net while enjoying private profit opportunities (Borys and Jemison, 1989: 235). What this means in practice could be well observed in the case of the CDM centers: their staff were on the public payroll, they could use public information and communication channels, and they could rely on public consultancy assignments in times of poor market demand. Hybrids certainly also pay a price for their exceptional status: they lose their autonomy – or, more precisely in the case of transitional economies, they have yet to gain autonomy. Neither collective nor state-owned enterprises in China are privatized overnight. Instead, they are gradually reorganized into a different ownership arrangement, which usually first means co-ownership by local governments and private operators. Thus, in the process of economic transition, organizational ownership and status are negotiable (Saich, 2000: 133). Hybrid constellations are thus a very common outcome of negotiations, because they represent a favorable compromise both for the former state owner who wants to retain political influence, and for the future private owner who is interested in turning the organization into a profitable enterprise but wishes to retain the local government as a patron. In transitional economies, hybrids can therefore be understood as a third organizational form, representing a governance structure that is discrete from the forms of both 'market' and 'hierarchy' (Nee, 1992:7).

Summary: Hybrids ≠ Hybrids

The literature review has revealed that dispensing with the private–public dichotomy and speaking of 'hybrids' adds little to the debate on new forms of governance, because hybrids differ widely in their nature depending on their context. In the OECD world of modern nation-states, 'hybrids' have the positive connotation of being actors that deliver governance outputs more efficiently than would be possible for public actors on their own. In the countries with transitional economies, 'hybrid' actors have the rather negative connotation of being a amalgam between state and non-state actors, neither particularly accountable nor effective in their governance performance. In non-OECD countries that are not transitional economies, but countries facing 'only' normal development challenges, the emergence and performance of 'hybrids' can be expected to take different forms again. Not only is the abolition of the public–private dichotomy when conducting research on hybrid actors in non-OECD areas long overdue: it is crucial that the differentiation of the contexts in which hybrids occur be made clear (since, for example, governance actors in a transitional economy like China cannot be compared to governance actors in straight market economies like Taiwan). Regional particularities might also be relevant distinguishing criteria; the debate on the East Asian economic miracles contains some valid points, with its emphasis on the cultural determinants of Asian governance styles, which might also represent significant variables for explaining the variation in the emergence and performance of hybrid actors (Cheung, 2000).

Explaining the performance of hybrid actors

Hybrid actors such as the public service units have been officially supported in China in order to increase the quality of the provision of public goods while being able to cut down on administrative spending. This institutional reform was also meant to improve the capacity of China's environmental governance system (Lo et al., 2001b: 55). Empirical research on the effectiveness of the provincial CDM centers, however, has shown only mixed results for their performance: they tend to be good at providing CDM project development services, but they underperform in their tasks related to the provision of public goods, such as information and skills.

The weak performance of CDM centers in the provision of public goods for CDM diffusion and CDM market development is unfortunately no exception to the rule, but can also be observed for other kinds of hybrid actors in China (Caulfield, 2006: 254; Lo et al., 2001b: 55). Previous research on other kinds of hybrid actor in China also came to the conclusion that the role and performance of environmental public service units could at best be described as 'mixed' (Lo et al., 2001b: 61). Although they would indeed have contributed to streamlining and downsizing the government's administrative apparatus, they would be preoccupied with income-earning activities because of the administration's inability to control them. Due to these experiences, '[r]eforming the service organizations has gradually emerged as the next frontier of institutional reform in China's environmental governance system' (Lo et al., 2001b: 61).

The expected solution to a governance problem has thus created a new governance challenge. But why do the hybrid actors show such mixed results in performance?

Hybrid actors have hybrid roles and a potential conflict of interests

The problem of assessing the performance of hybrid actors begins with defining what they are supposed to deliver. One reason for their mixed performance is that their objectives are already contradictory in themselves. Quite naturally, hybrid actors have hybrid roles. A fairly useful metaphor for understanding the role of hybrid governance actors has been their description as representing a 'hybrid being' – a centaur, for example (Schuppert, 2008: 11). Using the centaur as a metaphor for hybrid actors, the human features could be thought to represent their private business character, while the equine body could represent the 'sovereign state-steed' (Schuppert, 2008: 11). Of course, having two souls in one body is prone to conflict. This certainly also applies to hybrids like the CDM centers, which are meant to provide the public goods of information and skills in order to contribute to CDM diffusion and market development, but instead often choose to focus on their profit-making project-development services. CDM centers are thus typical examples of the dual character of public service units, which are simultaneously representative of their organization's corporate interests and agents of the state (Francis, 2001: 279).

Another obvious explanation of their mixed performance is their structural conflict of interest, partly due to the variety of their roles and mandates, described above. Given, on the one hand, the close link between hybrid actors and their state counterparts, and, on the other, the lack of clarity in their mandate, the coherence of their services may be undermined (Lo et al., 2001b: 61). This conflict of interest is not only inherent in Chinese hybrid organizations, but has also been a cause of mischief for many public enterprises in OECD countries that are supposed to run as performance-based institutions while providing unprofitable public goods and services; thus, they are supposed to optimize their behavior in terms of both adopting a market-based approach and correcting market failures (Weinert, 2001: 37; Stölting, 2001: 18). But optimization of behavior in relation to both of those objectives is not possible; one objective naturally has to be subverted by the other.

What is eventually translated into action are the interests of hybrids that are internally convergent. Again, CDM centers are a good example of this hypothesis. At least in relation to the development of CDM projects, CDM centers as profit-oriented entities and their affiliated government units share the same interests, albeit for slightly different reasons. CDM projects generate income for both of them: profits from project-development services for the CDM centers, and tax revenues from energy and industrial projects for the provincial government. Thus, provincial governments in China usually support the CDM because it offers them new sources of revenue. This entrepreneurial dimension of local governments has been identified in other areas before (Oi, 1996; Walder, 1995), and also in relation to the CDM: 'local governments in China act very much like profit-seeking businesses and [the] CDM provides a market for profit...It may be fair to say that financial gain

and technology transfer are two primary motivations of local governments for developing CDM projects' (Qi et al., 2008: 12). Besides their
own entrepreneurial motivation, local governments are also obligated
to advocate the non-profit-oriented demands and requirements made
on them by the central government, although they sometimes do this
only half-heartedly. The analysis of the constellation of interests within
CDM centers reveals CDM project development as a common denomi-
nator, with high incentives for both sides. Since interests converge in
the area of CDM project development, CDM centers enthusiastically
implement such activities.

Summarizing the discussion of the reasons for the mixed performance
of CDM centers and other public service units in China, the inherent
conflict of interests of hybrid actors seems to be the central explanation. The coherence of interests between their public half and private
dimensions on profit-making (also labeled 'bureaucratic entrepreneur-
ism' in Gore, 1998) guides their performance more than the underlying
interest of the public dimension to engage the hybrid organization in
the provision of public goods and services. The state is neither willing
(because its bureaucrats favor individual utility-maximization) nor able
(because state capacity and financial means are low at the local level) to
enforce its public interest.

Hybrid actors are growing out of control

Another reason for the mixed performance of hybrid actors is their quest
for financial independence from their affiliated governmental unit.
The consequence of that quest is the priority given by hybrid actors to
profit-making activities, instead of to providing services to their governmental sponsors. With growing financial independence, the hybrid
actor also gains decision-making autonomy. Although public patrons
still retain official authority over hybrids, there is a trend for hybrids to
grow out of control – a tendency that is not limited to China. In OECD
countries too, the supervision of public agencies reflects a fragile balance between granting necessary autonomy and exercising control over
service-delivery (Majone, 2005). The phenomenon of agencies becoming too independent from their public patrons has also been observed
for public agencies in OECD countries, as a consequence of political
decentralization and fragmentation (Christensen and Lægreid, 2001).

In China, the government administration has neither the capacity
nor the will to coerce public service units to deliver on their public
mandate. This applies to central as well as to provincial government,
albeit for different reasons. Decentralization has been a reform measure

designed to remedy the excessive concentration of economic and mana-gerial power in the central government, and to provide incentives to local government for efficient governance. Because decentralization has also included fiscal reforms favoring local government, central govern-ment lost several levers of control over what happens at the provin-cial level. The economic transition process has amplified this situation, because many tasks that used to be performed by government entities are now in the hands and under the control of private and hybrid enti-ties. As a consequence, China has experienced a rise in neo-localism, whereby provinces started to adhere only to local priorities and cater to local needs (Jia and Lin, 1994; Nee, 1992: 3). This included a form of corporatism between the provincial government and local companies which increasingly escaped central government control. As a conse-quence, the central government initiated a re-centralization campaign in the late 1990s, stripping local governments of fiscal revenues. The policy of outsourcing services to public service units has been one result of the limiting of local government budgets. With the erosion of central government supervision and the lack of financial means, the provincial governments' control of the hybrid public service units has become weaker. As I have noted, the increasing financial independ-ence of hybrid actors has also contributed to their autonomy. With the marketization of the Chinese economic system, the public service units now find it more profitable to sell their services to private consumers and private companies than to rely solely on their public patrons' finan-cial rewards. But other kinds of carrots and sticks still exist, and are responsible for the continuing government dependence of hybrid enti-ties. These include, for example, methods of political pressure and per-sonal reward exercised by the Communist Party in its cadre-evaluation system (Guo, 2007).

Ownership and autonomy determine hybrid actors' performance

Two main variables seem to be causal in the effectiveness of hybrid actors: the extent of their internal conflicts of interest, and the degree of control by their public patrons over the delivery of public goods. These variables, however, seem only to be secondary effects of changes in ownership rights and of a blurring of decision-making structures. If one takes a close look at the CDM centers, they reveal that performance does indeed vary with ownership structure and degree of autonomy in decision-making (see Table 6.1). None of the centers is subject to clear public or private ownership; instead there are public and private flows of money in various constellations that cover the costs of their establishment, staffing, and operating costs. In addition, the provincial government can even become an official shareholder if the center takes the form of a public company, as in Hunan. Centers also show great variation in their decision-making structures. Public agencies like the centers in Gansu and Yunnan are closely tied by the provincial governments' instructions, but have freedom of managerial decision-making on how to reach predefined targets. The more privately constituted centers in

Table 6.1   Possible hybrid models for CDM centers and their effectiveness

Form	Ownership	Autonomy	Example	Performance
Part of provincial government, but unit is allowed to make profit from selling services	Public: government funded establishment and pays operational costs	Low: the provincial government has directive power in decision-making	Gansu, Yunnan	Medium for private goods provision (project-development services); medium for public goods provision (these are provided, but quality is low)
'One entity with two heads,' one being part of government, the other being a private company	Public and private: government funded establishment and pays staff salaries; other operating costs met by self-financing	High: all decisions are made by the center's director, but he is obligated to incorporate 'public tasks'	Ningxia	Good for private goods provision (project development services); poor for public goods provision (information and training only during period of establishment)
Private company with government affiliation	Private: but government funded establishment costs and holds stake in the company; operating costs are self-financed	High: all decisions are made by the center's director, but he is obligated to incorporate 'public tasks'	Hunan	Good for private goods provision (project development services); poor for public goods provision (information and training only during period of establishment)

Ningxia and Hunan have almost independent decision-making power (especially in the case of their 'pure' private consultancy branches), but they try to cater to the governments' proposed tasks as much as possible, in order to retain political and financial support. The empirical findings show that, all other factors being equal, the more privately a hybrid is constituted and the more it has autonomy in its decision-making, the less it will provide public goods and the more it will focus on private goods provision.

Hybrid actors mirror the changing role of the state at the systemic level

What can now be learned from this discussion on the emergence and performance of hybrid actors concerning the state's role in the market? This section advances the argument that, echoing the analysis of the hybridization of public and private actors, an examination of the systemic level in China also reveals a blurring of the boundaries between state and market.

The state as a facilitating actor...

The state was always the central actor in China. Because state-centered governance has been a Chinese phenomenon for now over 3,500 years, many Chinese nationalists argue that authoritarian government is a natural feature of Chinese society (Zheng, 1999: 44). While a public–private dichotomy was almost nonexistent in Communist times, because almost no market existed under the planned economy, the era of reform that began in 1978 saw an incremental withdrawal of the state. For the first two decades, the role of the state in the Chinese economic miracle fitted well with the conception of the developmental state (Oi, 1992 and 1996; Wade, 1990; Cai, 2008). The state was strong, and intervened frequently in market decisions (by setting prices and restricting imports, for example), but was no longer omnipresent, and allowed the emergence of market institutions and market actors.

...turns into an entrepreneurial state

In the late 1990s, the role of the state switched once more: instead of being external market regulators and controllers, Chinese state agencies increasingly became entrepreneurial entities themselves (Walder, 1995; Oi, 1996; Tenev et al., 2002; Bruun, 1995; Yu, 1997). This switch began at the local level, where corporatism had always been strong, but 'trickled up' towards the central level, where even ministries now have their own profit-making private spin-offs. Consequently, China-watchers now speak no longer of the Communist Party being the country's government, but see it instead as resembling the managing board of the China Corporation (Fishman, 2005). While Western observers are usually only aware of the spectacular worldwide shopping trips of the China Investment Corporation (CIC), only little is known about the entrepreneurial behavior of state entities in China itself. Research on the role of CDM centers as an example of state entrepreneurialism has shed some light on this topic. With the state becoming an entrepreneur itself, the classical boundaries between state and market are no longer salient: the state itself has become a hybrid.

7
Conclusion: How to Make Hybrid Actors Accountable for the Provision of Public Goods

Underperformance of hybrid actors in the provision of public goods

The empirical results of this book reveal that the CDM centers in China show only mixed performance when assessed for their effectiveness in reaching the goals for which they have been established. These findings are confirmed by a review of the experiences of other types of hybrid actor in China's transitional economy (for example, hybrid research institutions and environmental protection bureaus) which tend to be good in the delivery of private goods, but weak in the delivery of public goods (Lam and Perry, 2001; Lo and Tang, 2007: 61). The Chinese provincial CDM centers are thus examples for hybrid agencies that – to put it bluntly – carry out some essential provision of public goods, but focus mainly on profit-making activities, so that their political mandate to provide public goods is neglected.

Hybrid actors, such as provincial CDM centers, tend to show good performance in the provision of public goods – such as CDM information and capacity in nascent markets – if their inherent conflict of interest can be overcome either by establishing mutually reinforcing incentives at their management level, or through external control by the state and/or civil society. These findings on the hybrid character of CDM centers fit with the argument that many hybrid actors in Chinese environmental governance only mirror the changing role of the state in becoming entrepreneurial institutions. In contrast to Western conceptions of China as a prototype of the East Asian developmental state (White and Wade, 1988; Wade, 1990), the policy experiments with

CDM centers at the local level confirm research that instead sees China as being in the midst of a transition towards becoming an entrepreneurial state (Yu, 1997; Duckett, 2004); this is also true in the realm of environmental governance (Lo and Tang, 2006), and for the CDM (Qi et al., 2008: 394). One consequence of this marketization and hybridization is a shift from providing public goods to focusing on the provision of private goods. In the face of a looming environmental catastrophe in China and a controversial Chinese response to the global effort to curb climate change, the crucial question must be asked: 'Who is left to serve the public interest?' It is argued that hybrid actors such as CDM centers lack accountability, because the local state lacks both the will and the capacity to take up its regulatory function, while civil society actors are few and also unable to act as watchdogs overseeing mighty semi-public actors within an authoritarian regime. Thus, possible options need to be discussed for symbiotic co-governance and the tricky question of how to keep hybrid actors in control.

Is symbiotic co-governance possible?

So far the performance of the new hybrid actors in Chinese environmental governance has been satisfactory in the provision of private goods, but disappointing in the provision of public goods. Contrary to the New Public Management paradigm, the outsourcing of public service provision has not yet been successful in China's transitional economy, where boundaries between the public and the private remain blurred (Chu, 2010). Is there nevertheless hope for successful co-governance of public, private, and even hybrid actors? The concept of co-governance has so far been used mainly to describe 'organized forms of interactions for governing purposes' (Kooiman, 2003: 97), and refers in practice to various forms of collaboration, cooperation, and co-management of private and public actors. Many empirical examples of such symbiotic, mutually reinforcing governance arrangements involving the public and the private sector have been observed in various regions and sectors (for Latin America, see Campbell and Fuhr, 2004; for the health and education sector, Van der Gaag, 1995). This evidence shows that public and private actors can complement each other in the provision of public goods.

What would co-governance look like for hybrid actors? It must be assumed that conditions conducive to co-governance need to be transferred from the realm of interaction between two actors (public and private) to the realm of a hybrid actor's internal operations (which might be the management level of strategic decision-making). Just as incentives for public and private actors have to be in the right constellation at the level of their interaction for cooperation to occur, incentives are required for decisions taken at the managerial level of hybrid organizations. As I have noted, however, the incentives found in the hybrid organizations in China are not mutually reinforcing, but unfortunately push in different directions: mandates, on the one hand, demand the provision of public goods that are by definition free for everyone, and on the other hand demand that the organization be able to finance itself by profit-generating activities. While the public mandate is backed by career rewards, the private mandate is incentivized by monetary rewards. But because the latter mandate is also conducive to advancing a career within a prospering public agency, and brings in revenues for its public patron, decision-makers in public agencies tend to pursue monetary incentives first, and only superficially cater to their public tasks.

Their focus on profit-maximizing activities can be explained and recognized as appropriate because they are what they are – namely, hybrid actors. If they were purely public bureaucracies with a clear mandate for the provision of public goods, their tendency to concentrate on profit-making activities could be labeled as rent-seeking. But since rent-seeking is defined as 'any manipulation of the law or of government authority in order to generate or appropriate an economic rent' (World Bank, 1996: viii), it does not apply to the CDM centers. In concentrating their work on for-profit project development, they do not bend the law or abuse government authority. Their profit-focused behavior is instead legitimate, because they are partly private companies. Thus, although a focus on the provision of public goods would be desirable, the lack of appropriate incentives to do so puts the blame not on the actors, but on the structures within which they operate.

Nevertheless, critical note must be registered on the opportunities for the stripping of government assets that arise due to the prevalent incentive structures for hybrid actors in countries in transition. The coexistence of different institutional formations, with the hybridization of their actors, has created opportunities for persons heading such hybrid organizations to turn assets of their 'governmental part' into profits for the 'private part' (Lo et al., 2001: 61; Tenev and Zhang 2002: 22; Caulfield 2006). But if 'profits for the private part' in effect become personal profits, this can be considered a form of rent-seeking (for a classical definition see Tullock, 1967; for a more recent overview, see Tullock, 1993). On the other hand, although the CDM centers can take advantage of their public counterparts by, for example, using their information channels and exploiting their political status, this produces no disadvantages for the provincial government at all: on the contrary, the effective use of their government affiliation is also advantageous for the government if CDM diffusion by the centers leads to additional sources of revenue.

The consequence of these findings on hybrid actors in China should be to shift the research focus from actor constellations to incentive constellations. Research and debate are also required on what kind of criteria could be used to evaluate the performance of hybrid actors in a normative way. Thus the main focus should not be on the public versus the private, or the state versus the market, but on the question of how to create incentive structures that give value to public goods and that would enable symbiotic co-governance. This objective is in line with Walder's conclusion about the effects of the transitional nature of China's economy:

> [The] most important conclusion [is that] a transitional economy must alter incentives not merely for individuals and firms but for government agencies and government officials themselves, for the behavior of the latter can have enormous economic consequences ... the task is not to revile state involvement but to change it. (Walder, 1996: 16)

Who is left to check on the system?

Creating an incentive structure conducive to the provision of public goods is of course a major challenge. There may be limitations on the extent to which monetary incentives can be provided for the provision of public goods, especially for financially weak developing countries. If the carrots are few, what about the sticks? Are there coercive means available to ensure the provision of public goods?

Although governance can take place without government playing an active role, the existence at least of a 'shadow of hierarchy' is assumed to be necessary to achieve it (Börzel, 2007); thus, the state must at least give a credible impression that it would be able and willing to enforce the provision of public goods by hybrid actors. This precondition is often called the 'governance paradox,' whereby sufficient state capacity needs to exist for an effective outsourcing of governance functions to private (or hybrid) actors – although this private governance was thought to be a supplement for ineffective state governance in the first place (Börzel, 2008: 118).

At first sight, China might seem the ideal place for realizing new private governance arrangements, if that depended chiefly on coercive state power. China has a strong central state with rule-enforcement power. The government is interested in experimenting with market mechanisms and with the delivery of public goods by private actors. The central government is also interested in streamlining its bureaucracies, cutting government expenditure, and experimenting with recommendations from 'new public management.' Are these indeed ideal conditions?

Unfortunately, even in China state capacity to control for the provision of public goods by non-state actors is not as strong as many imagine. Krugmann even speaks of the Chinese government as being 'too soft,' because it adheres to a model of 'crony capitalism' in which corporatist relationships between state and no-state actors are too strong (Krugman, 1994). Indeed, local state control, at least, is often lax, maybe more because of a lack of motivation and interest in the provision of public goods than due to a lack of capacity. Private and hybrid actors that have taken up mandates for the provision of public goods are thus growing out of control – and are not delivering as expected.

The participation of hybrid actors in governance makes the situation even more complicated. The public–private dichotomy of actors in the analysis of OECD countries at least provided some conceptual framework for who would control whom in the provision of public goods. With hybrids, not only do the boundaries between public and private actors become blurred – so too does that between the subject and object of control. The question with special relevance for the governance performance of hybrid actors is: Who is going to exercise this shadow of hierarchy if hybrid actors are the objects under it (their private half providing governance outputs) as well its subject (their public half demanding governance provision). In this schizophrenic constellation of interests, it is usually the convergent interest in maximizing private utilities that gains the upper hand, at the expense of the provision of public goods.

If the public interest is nevertheless to be served, this might be achieved by someone external to the hybrid actor taking up the controlling role in the provision of public goods. Two fundamental trajectories are possible – producing a stand-alone solution or a combined approach. One possible strategy is to strengthen state capacities to enable public actors to exercise their full powers of control over the provision of public goods. A second strategy would be to concentrate instead on strengthening non-state actors within civil society, so that they can adopt a controlling function and hold the state, as well as hybrid and private actors, accountable for their performance.

The first strategy – that of fostering state capacity – remains, of course, part of the meaningless endeavor of strengthening state capacity for the sake of improved state capacity. In a country like China, the recommendation to strengthen state capacity in pursuit of better control over governance outputs is too simplistic, because even in its authoritarian system there is no single state actor. Instead, there is a multitude of levels of the state in China, comprising various ministries and agencies on the horizontal level, and a variety of central-, provincial-, district-, county-, city-, township- and village-level state actors vertically. As a means of strengthening state capacity in China, the disaggregation of the state into its components at the different levels is long overdue, because each level of government faces different barriers and incentives in improving its performance (Oi, 1995: 1,147). Research on state capacity in China usually arrives at the conclusion that it is at the local level that will and capacity are lacking, while the central level generates good macroeconomic policies but is weak in enabling their implementation. Both levels, however, are afflicted with governance deficits in terms of accountability, transparency, and legitimacy. But even if the Chinese government resembled a 'modern' and democratic state of the OECD type, its ability to supervise the private (or hybrid) provision of public goods effectively might not be taken for granted, because a government is not an 'exogenous and neutral arbiter' that steers the economic system from outside, but can instead be understood as an 'endogenous element of the system with the same information and incentive constraints as other economic agents in the system' (Aoki et al., 1997: xvii).

The second strategy of looking to non-state actors to take on watchdog and control functions in relation to the governance performance of hybrid and private actors at least has the advantage of targeting not-for-profit actors who (usually) pursue no economic interests (on the emergence of 'entrepreneurial NGOs' see Young, 1986, and Meyer, 1995). Their incentive constellation should be more in line with the public interest, as NGOs – at least in principle – directly represent the interests of civil society groups. NGOs can thus also compensate for some of the deficient accountability of hybrid actors. Addressing and strengthening civil society, and involving it constructively in governance arrangements, are thus popular proposals for overcoming accountability deficits (Haas, 2004; Newell, 2008; Mason, 2005). Such deficits on the part of public administrations have already been identified by Lipsky in his book on the 'freedom' or 'arbitrariness' of street-level bureaucrats (Lipsky, 1983); and accountability deficits are certainly a given for hybrid actors in China.

While these proposals make sense for democracies with well-developed civil societies, there is a lack of research on how to include non-governmental organizations in governance arrangements within authoritarian systems. The common explanation for this research gap is that there are no NGOs in those countries, or that those NGOs that do exist are in any case not allowed to participate in governance (White et al., 1996). But such an explanation simplifies the situation. Although, for example, NGOs comparable to Western advocacy organizations are indeed few in China, there are many other non-profit organizations that are somewhere on the continuum between public and private actors. There are many kinds of hybrid 'NGOs' that do not fit Western categories: for example, GONGOs (government-owned non-governmental organizations) and QuaNGOs (quasi-governmental non-governmental organizations) (see Table 7.1). While GONGOs are clearly government-owned and controlled, QuaNGOs are privately owned organizations that have a government affiliation (for an attempt to define 'QuaNGO,' see Greve et al., 1999; for their situation in China, see Saich, 2000). QuaNGOs are therefore the hybrids of the not-for-profit sector, comparable to the hybrid agencies of the for-profit sector.

Outsourcing the provision of public goods to QuaNGOs has the advantage that profit-maximizing motives are less in competition with the provision of public goods. On the other hand, making non-governmental organizations fully responsible for public service provision

Table 7.1 Potential providers of public services: ownership and incentives

	Public	**Hybrids**	**Private**
For-profit	State-owned enterprises, including public service units of a for-profit nature	*Shiye danwei,* which are both privately and publicly owned	Private enterprises
Not-for-profit	Government departments, government-owned non-profits (GONGOs)	Quasi-governmental NGOs (QuaNGOs)	Privately owned nonprofits (NGOs)

Source: Adapted from Wong, 2004: 21.

does not solve the accountability problem either, because NGOs tend to be accountable only for the special interest group they represent, and not for the entire public. Moreover, many countries that have followed this suggestion and that play host to social organizations providing public services also experience a growing problem of legitimacy for the state government (on state–NGO competition, see Gordenker and Weiss, 1995: 551). Thus, instead of advocating a new 'business–civil society' dichotomy, one might follow Katherine Morton, who proposes an all-encompassing approach reconciling economic incentives with participatory approaches based on experiences with environmental management reforms in China (Morton, 2005: 4).

Apparently there is no 'one size fits all' approach to the provision of public goods and services. As David Nee argued as far back as 1992, there is no ideal organizational arrangement:

> [N]o general organizational form is best suited for all institutional environments. Formalized and hierarchical forms of classical state socialism are well suited to a homogeneous and stable environment. But during periods of rapid change and institutional uncertainty, organizational forms that are more flexible, informal, and open to entrepreneurship exhibit superior adaptive capacity ... (Nee, 1992: 4)

The emergence of hybrid actors in transitional economies and their status as relics in OECD countries has opened up new possibilities for governance constellations whose merits can be tested in their particular circumstances.

The inclusion of hybrids in research on governance may transgress traditional analytical boundaries, but it contributes important new categories of actor into the discussion – for example, agencies, 'professional' NGOs, and QuaNGOs (see Figure 7.1), which are better able to reflect today's complex realities than the traditional public–private dichotomy, or the newer 'state–market–community' triangle (Wolf, 2008; Salamon, 1987).

Are hybrid actors a transitional or enduring phenomenon?

Are hybrid actors worth the widespread attention they receive? Or are they, rather, simply a transitional phenomenon that will dissolve with time, so that we can eventually return to a reliance on the public–private dichotomy that has proved its merits in explaining many governance

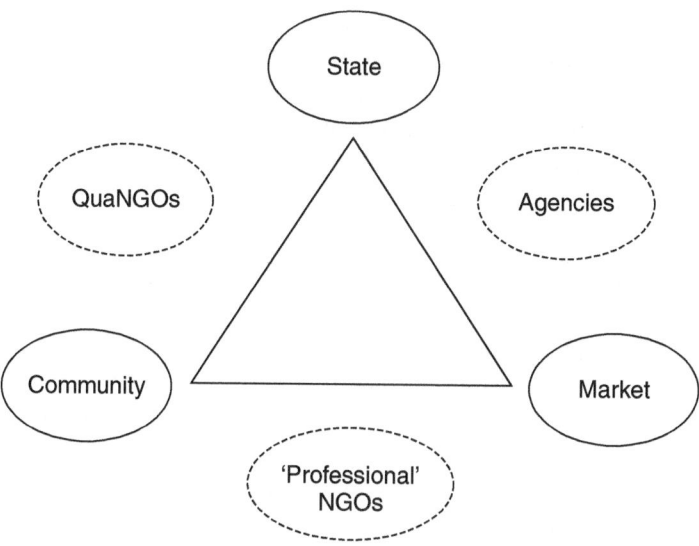

Figure 7.1 Traditional and hybrid actor types in environmental governance

arrangements and outcomes? Should our expectations concerning their future be the same for hybrids in OECD countries as for hybrids in transitional economies? For the latter type of hybrid, we already have an intuition that their existence represents only one step in their country's evolution from a planned towards a market economy. This intuition is based on notions of a path-dependency towards the 'modern' and democratic nation-state of the OECD type, which at least used to be characterized by a clear division between the public and the private. But the emergence of hybrid actors in OECD countries should prompt some doubts about what future governance models are going to look like. Will hybrids become the norm, or are they merely transitional phenomena? Are they transitional phenomena only in countries like China, or for all countries and areas of limited and not-so-limited statehood? Are hybrids in China a sustainable trend leading towards a new 'entrepreneurial–corporatist' variety of Chinese governance model?

There is already a lively discussion on the shifting role of the state – from planner, to facilitator, to emerging entrepreneur (Yu, 1997; Oi, 1998; Duckett, 2004). A governance model having the 'entrepreneurial state' at its center while acknowledging a strong element of corporatism might indeed become a specifically Chinese variety of governance. But if one takes a critical look at the changing role of (especially municipal) governments in OECD countries, which – for the same reasons of fiscal constraint – have increasingly become entrepreneurial rather than regulatory in their nature, one wonders whether the entrepreneurial state model might not become the future model of governance world-wide. If it does, much more thought will need to be put into the question of how to ensure the provision of public goods, whether by public, private, or hybrid actors, if those actors are entrepreneurial and profit-orientated in their nature.

Challenges to future climate governance in China

How does the CDM experience in general – and not just the story of the CDM centers – influence environmental governance in China? One thing that the CDM clearly has achieved, despite all criticism, is a change in attitudes. Many representatives of Chinese companies stated in interviews that they now perceived climate-protection measures as positive – partly because they were for the global good, and because the central government was eager to enforce energy-saving policies; but in truth because they enabled money to be made. While traditional command-and-control instruments – for example, those for the reduction of SO_2 emissions – were barely complied with, and often induced corrupt practices, market mechanisms like the CDM at least offer the prospect of revenues, although market mechanisms also leave much scope for the rules to be bent – for example, those on additionality. As in many other countries, attaching a monetary value to something formerly regarded as an externality dictates new behavioral incentives for actors involved in environmental governance. Judged upon the success of the Chinese CDM market, these incentives seem to be sufficient to encourage the assumption of the transaction costs (for example, that of obtaining information) and the risks to private actors of getting involved in the CDM business. The policy experiment of establishing provincial CDM centers further contributed to the tapping of China's vast CDM potential – especially that of its western provinces. Consequently, the Chinese delegates to the climate negotiations have enthusiastically advocated the CDM, and wish to see its continuation and even expansion in the post-2012 commitment period.

The CDM has become a new business opportunity and a global market with its own dynamics. In China, it has enhanced tendencies towards marketization of environmental governance, and towards a hybridization of the actors involved. Public agencies, even government ministries, compete with each other to win their piece of the CDM cake. State actors thus become entrepreneurs. But on the international level, too, the CDM generates the same trend of state actors becoming entrepreneurs and hybrid actors. Governments of Annex I countries of the Kyoto Protocol have a dual role in the CDM: as members of the Kyoto Protocol they can act as regulators and set the rules of the game; and as national governments they can use CERs for their and/or their national companies' GHG emission-reduction obligations, and consequentially appear as buyers on the market. A hybrid nature with partly contrasting incentives is thus also detectable among non-Chinese state actors. Much like the CDM center hybrids in their motivation, some foreign governments seemed to have become active in CDM capacity-development programs in China more for their own benefit than in response to local mitigation and adaption needs. They act primarily as facilitators for CER purchases, and also compete with other donors for pieces of the cake – in their case being able to support provinces with good CDM potential, in which they can claim the 'first right of CER purchase.'

There is one finding from the empirical research on the CDM that has profound implications for future governance: all actors involved in the CDM in China – companies, state actors, research institutions, even many NGOs – welcome the CDM as a new business opportunity. If all converge in their in interests and become CDM entrepreneurs, who is then left to police the integrity of the mechanism? If even state actors and NGOs become entrepreneurs, much more thought must be put into the question of how the diversity of public, private, and hybrid actors can be integrated into a governance system that is also supposed to deliver public goods.

Notes

2 The Need for Capacity Development in the Early CDM Market

1. HFC and PFC projects reduce the greenhouse gas fluoroform in manufacturing processes of certain refrigerants. This project type is heavily criticized because these projects reap a high CER profit due to the high global warming potential of fluoroform, while abatement costs are comparatively low. Some project owners are accused of running refrigerant production processes that can be turned into CDM projects solely in order to earn CER revenues (CDM Watch, 2010).
2. In the subsequent text, the term 'province' is used to describe all provincial-level units, including the four centrally administered municipalities (Beijing, Shanghai, Tianjin, and Chongqing) and the five autonomous regions (Guangxi, Inner Mongolia, Ningxia, Tibet, and Xinjiang).

4 A Case Study on the Performance of Four CDM Centers

1. *Guanxi* can be translated positively as 'networks' and negatively as 'nepotism.' See Jacobs, 1979; Park and Luo, 2001.

References

ADB (2001), *ADB Recommends Market-Based Instruments to Help Protect the Environment in the People's Republic of China* (Manila: Asian Development Bank).

ADB (2006), *People's Republic of China: Establishment of the Clean Development Mechanism Fund*, Technical Assistance Report (Manila: Asian Development Bank).

Akerlof, G. A. (1970), 'The market for "lemons": quality uncertainty and the market mechanism,' *Quarterly Journal of Economics* 89, 488–500.

Anderson, S. T. and Newell, R. G. (2004), 'Information programs for technology adoption: the case of energy efficiency audits,' *Resource and Energy Economics* 26, 27–50.

Andonova, L., Betsill, M. and Bulkeley, H. (2007), 'Transnational climate change governance,' paper presented at the 2007 Amsterdam Conference on the Human Dimensions of Global Environmental Change, Amsterdam.

Aoki, M., Kim, H.-K. and Okuno-Fujiwara, M. (1997), 'Introduction,' in Aoki, M., Kim, H.-K. and Okuno-Fujiwara, M. (eds), *The Role of Government in East Asian Economic Development* (Oxford: Clarendon Press).

Argote, L. and Epple, D. (1990), 'Learning Curves in Manufacturing,' *Science* 247, 920–4.

Arrow, K., Bolin, B., Costanza, R., Dasgupta, P., Folke, C., Holling, C. S., Jansson, B.-O., Levin, S., Mäler, K.-G., Perrings, C. and Pimpentel, D. (1995), 'Economic growth, carrying capacity, and the environment,' *Ecological Economics* 15, 91–5.

Arrow, K. J. (1962), 'The Economic Implications of Learning by Doing,' *Review of Economic Studies* 29, 155–73.

Assunção, P. T. B., Fuhr, H. and Späth, B. (1993), *Internationale Organisationen, Entwicklungsverwaltungen und Kleingewerbeförderung in der Dritten Welt* (Baden-Baden: Nomos)

Babu, N. Y. D. and Michaelowa, A. (2003), *Removing Barriers for Renewable Energy CDM Projects in India and Building Capacity at the State Level*, HWWA Report (Hamburg: Hamburg Institute of International Economics).

Badian, E. (1997), *Zöllner und Sünder. Unternehmer im Dienst der römischen Republik* (Darmstadt: Wissenschaftliche Buchgesellschaft).

Barney, J. B. and Hansen, M. H. (1994), 'Trustworthiness as a Source of Comparative Advantage,' *Strategic Management Journal* 15, 175–90.

Barreto, L. (2003), 'Gaps and Needs in Technology Diffusion Models: The Perspective of an Energy-systems Modeler,' Paper presented at the Workshop on Clean Technologies Diffusion Modeling, IPTS, Seville.

Bechert, S. (1995), *Die VR China in internationalen Umweltregimen* (Münster, LIT Verlag).

Biermann, F. and Pattberg, P. (2008), 'Global Environmental Governance: Taking Stock, Moving Forward,' *Annual Review of Environment and Resources* 33, 277–94.

Bilderbeek, R., Hertog, P. D., Marklund, G. and Miles, I. (1998), *Services in Innovation: Knowledge-Intensive Business Services (KIBS) as Co-producers of Innovation*, Services in Innovation and Innovation in Services (SI4S), project synthesis paper, Luxembourg.

Borys, B. and Jemison, D. B. (1989), 'Hybrid arrangements as strategic alliances: Theoretical issues in organizational combinations,' *Academy of Management Review* 14, 234–49.

Börzel, T. A. (2007), 'Regieren ohne den Schatten der Hierarchie. Ein modernisierungstheoretischer Fehlschluss?,' in Risse, T. and Lehmkuhl, U. (eds), *Regieren ohne Staat? Governance in Räumen begrenzter Staatlichkeit* (Baden-Baden: Nomos).

Börzel, T. A. (2008), 'Der „Schatten der Hierarchie"– Ein Governance-Paradox?,' in Schuppert, G. F. and Zürn, M. (eds), *Governance in einer sich wandelnden Welt* (Wiesbaden: VS Verlag für Sozialwissenschaften).

Breslin, S. (ed.) (2009), *China and the Global Political Economy* (Basingstoke: Palgrave Macmillan).

Bruun, O. (1995), 'Political hierarchy and private entrepreneurship in a Chinese neighborhood,' in Walder, A. G. (ed.), *The waning of the communist state: Economic origins of political decline in China and Hungary* (Berkeley: University of California Press)

Bünte, M. (2006), 'Hybride Regime in Südostasien: Genese, Gestalt und Entwicklungsdynamiken,' in Pickel, G. and Pickel, S. (eds), *Demokratisierung im internationalen Vergleich. Neue Erkenntnisse und Perspektiven* (Wiesbaden: VS Verlag für Sozialwissenschaften).

Burgi, M. (1999), *Funktionale Privatisierung und Verwaltungshilfe. Staatsaufgabendogmatik – Phänomenologie – Verfassungsrecht* (Tübingen: Mohr).

Burns, J. P. (2003a), ' "Downsizing" the Chinese State: Government Retrenchment in the 1990s,' *China Quarterly* 175, 775–802.

Burns, J. P. (2003b), 'Governance and public sector reform in the People's Republic of China,' in Cheung, A. B. L. and Scott, I. (eds), *Governance and Public Sector Reform in Asia. Paradigm shifts or business as usual?* (London and New York: Routledge).

Cai, K. G. (ed.) (2008), *The Political Economy of East Asia. Regional and National Dimensions* (Basingstoke: Palgrave Macmillan).

Campbell, T. E. J. and Fuhr, H. (eds) (2004), *Leadership and innovation in subnational government: case studies from Latin America* (Washington, DC: World Bank).

Cashore, B., Auld, G. and Newsom, D. (2004), *Governing through Markets: Forest Certification and the Emergence of Non-State Authority* (New Haven: Yale University Press).

Castro, P. (2008), *Empirical analysis of performance of CDM projects: case study China*, Discussion Paper CDM-6 (London: Climate Strategies).

Caulfield, J. L. (2006), 'Local government reform in China: a rational actor perspective,' *International Review of Administrative Sciences* 72, 253–67.

CCICED (2001), *The Clean Development Mechanism and the Promotion of Sustainable Development in China's Western Regions: Policy Instruments for Cleaner Technology Transfer* (Beijing: China Council for International Cooperation on Environment and Development (CCICED)).

CDASED/World Bank (2001), *Business Development Services for Small Enterprises: Guiding Principles for Donor Intervention* (Washington, DC: Committee of Donor Agencies on Small Enterprise Development (CDASED)/World Bank).

CDM Watch (2010), *HFC-23 offsets in the context of the EU emission trading scheme*, at www.cdm-watch.org/wordpress/wp-content/uploads/2010/07/HFC-23_Policy-Briefing1.pdf.

Chan, H., Cheung, K. C. and Lo, J. M. K. (1993), 'Environmental control in the PRC,' in Nagel, S. and Mills, M. (eds), *Public Policy in China* (Westport, CT: Greenwood Press).

Chandler, A. (1977), *The Visible Hand* (Cambridge, MA: Harvard University Press).

Chen, Q., Chen, J. and Zhang, Q. (2002), 'On the Reform of Local Government Organizations in China,' *Chinese Public Administrative Review* 1: 2, 35–50.

Cheung, A. B. L. (2000), 'Globalization versus Asian Values: Alternative Paradigms in Understanding Governance and Administration,' *Asian Journal of Political Science* 8, 1–16.

Cheung, A. B. L. and Scott, I. (2003), 'Governance and public sector reforms in Asia: paradigms, paradoxes and dilemmas,' in Cheung, A. B. L. and Scott, I. (eds), *Governance and Public Sector Reform in Asia. Paradigm shifts or business as usual?* (London and New York: Routledge).

Chin, G. T. (ed.) (2010), China's Automotive Modernization. The Party-State and Multinational Corporations (Basingstoke: Palgrave Macmillan).

CHINA CSR.COM (2007), *Provincial CDM Service Centers Set Up In China*, at www.chinacsr.com/en/2007/09/20/1699-provincial-cdm-service-centers-set-up-in-china.

Chipman, K. and Morales, A. (2010), *China Turns Negotiating Tables on U.S. at Stalled Cancun Climate Meeting*, at www.bloomberg.com/news/2010-12-03/china-turns-negotiating-tables-on-u-s-at-stalled-cancun-climate-meeting.html.

Chou, K. P. B. (2004), 'Civil Service Reform in China, 1993–2001,' Paper presented at the Hong Kong Political Science Association Annual General Meeting, City University of Hong Kong.

Christensen, J. L. and Lundvall, B.-Å. (2004), *Product Innovation, Interactive Learning and Economic Performance* (Amsterdam: Elsevier).

Christensen, T. and Lægreid, P. (2001), 'New Public Management. The Effects of Contractualism and Devolution on Political Control,' *Public Management Review* 3:1, 73–94.

Chu, Y.-W. (ed.) (2010), *Chinese Capitalisms. Historical Emergence and Political Implications* (Basingstoke: Palgrave Macmillan).

Corell, E. and Betsill, M. M. (2001), 'A comparative look at NGO influence in international environmental negotiations: desertification and climate change,' *Global Environmental Politics*, 1: 4, 86–107.

Corfee-Morlot, J., Kamal-Chaoui, L., Donovan, M. G., Cochran, I., Robert, A. and Teasdale, P.-J. (2009), *Cities, Climate Change and Multilevel Governance*, OECD Environmental Working Papers No. 14 (Paris: OECD).

Cornes, R. and Sandler, T. (1996), *The Theory of Externalitities, Public Goods, and Club Goods* (Cambridge: Cambridge University Press).

Croissant, A. (2002), 'Einleitung: Demokratische Grauzonen – Konturen und Konzepte eines Forschungszweigs' in Bendel, P., Croissant, A., Rüb, F. W. (eds),

Zwischen Demokratie und Diktatur. Zur Konzeption und Empirie demokratischer Grauzonen (Opladen: Leske & Budrich).

Cutler, A. C., Haufler, V. and Porter, T. (eds) (1999), *Private authority and international affairs* (Albany, NY: State University of New York Press).

Del Río, P. (2007), 'Encouraging the implementation of small renewable electricity CDM projects: An economic analysis of different options,' *Renewable and Sustainable Energy Reviews* 11, 1,361–87.

Deng, X. (1983), *The Selected Works of Deng Xiaoping: 1975–1982* (Beijing: Foreign Language Press).

Deng, X. (1987), *Building Socialism with Chinese Characteristics [Jianshe Zhongguo Teshe de Shehui Zhuyi]* (Beijing: People Publishing House).

Ding, X. L. (1994), 'Institutional Amphibiousness and the Transition from Communism: The Case of China,' *British Journal of Political Science* 24, 293–318

Döhler, M. and Jann, W. (eds) (2007), *Agencies in Westeuropa* (Wiesbaden: VS Verlag für Sozialwissenschaften).

Dosi, G. (1988), 'The nature of the innovation process,' in Dosi, G., Freeman, C., Nelson, R., Silverberg, G. and Soete, L. (eds), *Technical Change and Economic Theory* (London: Frances Pinter).

Douthwaite, B., Keatinge, J. D. H. and Park, J. (2001), 'Why promising technologies fail: the neglected role of user innovation during adoption,' *Research Policy* 30, 819–36.

Draude, A. (2007), *How to Capture Non-Western Forms of Governance. In Favour of an Equivalence Functionalist Observation of Governance in Areas of Limited Statehood*, SFB-Governance Working Paper Series (Berlin, DFG Research Center (SFB), 700 Governance in Areas of Limited Statehood).

Duckett, J. (2004), 'The evolving institutional environment and China's state entrepreneurship,' in Krug, B. (ed.), *China's Rational Entrepreneurs. The development of the new private business sector* (London and New York: RoutledgeCurzon).

Dudek, D. J., Stewart, R. B. and Wiener, J. B. (1992), 'Environmental Policy for Eastern Europe: Technology-Based Versus Market-Based Approaches,' *Columbia Journal of Environmental Law* 17, 1–52.

Economy, E. (2007), 'Environmental Governance: the Emerging Economic Dimension,' in Mol, A. P. J. and Carter, N. T. (eds), *Environmental Governance in China* (Abingdon and New York: Routledge).

Economy, E. C. (2004), *The River Runs Black: the Environmental Challenge to China's Future* (Ithaca and London: Cornell University Press).

Edeling, T., Jann, W., Wagner, D. and Reichard, C. (eds) (2001), *Öffentliche Unternehmen. Entstaatlichung und Privatisierung?* (Opladen: Leske & Budrich).

Edin, M. (2003), 'State Capacity and Local Agent Control in China: CCP Cadre Management from a Township Perspective,' *China Quarterly* 173, 35–52.

Edquist, C. and Jacobsson, S. (1988), *Flexible Automation: The Global Diffusion of New Technology in the Engineering Industry* (Oxford: Basil Blackwell).

Edquist, C. (2005), 'Systems of Innovation. Perspectives and Challenges,' in Fagerberg, J., Mowery, D. C. and Nelson, R. R. (eds), *The Oxford Handbook of Innovation* (Oxford: Oxford University Press).

Eisinger, P. K. (1998), *The Rise of the Entrepreneurial State. State and Local Economic Development Policy in the United States* (Madison: University of Wisconsin Press).

Ellis, J. and Kamel, S. (2007), *Overcoming Barriers to Clean Development Mechanism Projects* (Paris: OECD).

Erdmann, G. (2002), 'Neopatrimoniale Herrschaft – oder: Warum es in Afrika so viele Hybridregime gibt,' in Bendel, P., Croissant, A., Rüb, F. W. (eds), *Zwischen Demokratie und Diktatur. Zur Konzeption und Empirie demokratischer Grauzonen* (Opladen: Leske & Budrich).

Erdmann, G. (2007), 'Demokratisierung in Afrika und das Problem hybrider Regime,' in Däubler-Gmelin, H., Münzing, E. and Walther, C. (ed.), *Afrika – Europas verkannter Nachbar* (Frankfurt a.M.: Peter Lang Verlag).

Fan, Q., Dong, Y. and Zeng, D. Z. (2009), 'Innovation, Competitiveness, and Economic Development,' in Fan, Q., Li, K., Dong, Y., Zeng, D. Z. and Peng, R. (eds), *Innovation for Development and the Role of Government. A perspective from the East Asia and Pacific Region* (Washington, DC: World Bank).

Findeis, A. (2007), *Technologie- und Gründerzentren als Instrument zur Förderung der Regionalentwicklung. Eine regionalwirtschaftliche Erfolgsanalyse unter Berücksichtigung der Gründungsforschung* (Hamburg: Verlag Dr. Kovac).

Firth, M., Lin, C., Liu, P. and Wong, S. M. L. (2009), 'Inside the black box: Bank credit allocation in China's private sector,' *Journal of Banking and Finance*, 1,144–55.

Fisher, A. and Rothkopf, M. (1989), 'Market Failure and Energy Policy: A Rationale for Selective Conservation,' *Energy Policy* 17, 397–406.

Fishman, T. C. (2005), *China, Inc.: How the Rise of the Next Superpower Challenges America and the World* (Scribner).

Flavin, C. and Stoke, L. (2008), *State of the World 2008: Ideas and Opportunities for Sustainable Economies* (Washington, DC: Worldwatch Institute).

Fliegel, F. C. and Kivlin, J. E. (1966), 'Attributes of innovations as factors in diffusion,' *American Journal of Sociology* 72, 235–48.

Foxon, T. J. (2003), *Inducing Innovation for a low-carbon future: drivers, barriers and policies* (London: The Carbon Trust).

Francis, C.-B. (2001), 'Quasi-Public, Quasi-Private Trends in Emerging Market Economies: The Case of China,' *Comparative Politics* 33, 275–94.

Fu, J. and Li, J. (2009), *China played 'constructive' role, Wen says*, at www.chinadaily.com.cn.

Fuhr, H. (1997), 'Institutional change and new incentive structures for development: Can Decentralization and better local governance help?' Paper presented at the Annual Meeting of the Section 'Developing countries' of the Verein für Sozialpolitik, Zürich, June 6–7.

Fukuda-Parr, S., Lopes, C. and Malik, K. (2002), 'Institutional innovations for capacity development,' in S. Fukuda-Parr, Lopes, C and Malik, K. (eds), *Capacity for Development: New Solutions to Old Problems* (London: Earthscan).

G77 and CHINA (2008), *Proposal by the G77 and China for a Technology Mechanism under the UNFCCC*, at unfccc.int/files/meetings/ad_hoc_working_groups/lca/application/pdf/technology_proposal_g77_8.pdf.

Garcelon, M. (1997), 'The Shadow of the Leviathan: Public and Private in Communist and Post-Communist Society,' in Weintraub, J. and Kumar, K. (eds), *Public and Private in Thought and Practice. Perspectives on a Grand Dichotomy* (Chicago and London: University of Chicago Press).

Golder, P. N. and Tellis, G. J. (1997), 'Will It Ever Fly? Modeling the Takeoff of Really New Consumer Durables,' *Marketing Science* 16, 256–70.

Goldmark, L. (1996), *Business Development Services: A framework for analysis* (Washington DC: Inter-American Development Bank).

Goldstein, S. M. (1995), 'China in Transition: The Political Foundations of Incremental Reform,' *China Quarterly* 144, 1,105–31.

Golove, W. H. and Eto, J. H. (1996), *Market Barriers to Energy Efficiency: A Critical Reappraisal of the Rationale for Public Policies to Promote Energy Efficiency* (Berkeley, CA: Lawrence Berkeley National Laboratory).

Gordenker, L. and Weiss, T. G. (1995), 'Pluralising global governance: Analytical approaches and dimensions,' *Third World Quarterly* 16, 357–87.

Gore, L. P. (1998), *Market communism: The institutional foundations of China's post-Mao Hypergrowth* (Hong Kong: Oxford University Press).

Greenberg, M. R. (2006), 'The Diffusion of Public Health Innovations,' *American Journal of Public Health* 96, 209–10.

Greve, C., Flinders, M. and Thiel, S. V. (1999), 'Quangos: What's in a Name? Defining Quangos from a Comparative Perspective,' *Governance* 12, 129–46.

Grieg-Gran, M., Porras, I. and Wunder, S. (2005), 'How Can Market Mechanisms for Forest Environmental Services Help the Poor? Preliminary Lessons from Latin America,' *World Development* 33, 1,511–27.

Grubb, M. (2004), 'Technology Innovation and Climate Change Policy: an overview of issues and options,' *Keio Journal of Economics* 41, 103–32.

Grübler, A., Nakicenovic, N. and Victor, D. G. (1999), 'Dynamics of energy technologies and global change,' *Energy Policy* 27, 247–80.

GTZ (2000), *Wirkungsmonitoring in Projekten der Institutionenentwicklung im Umweltbereich* (Eschborn: GTZ).

GTZ (2008), *CDM Highlights 61*, at www.gtz.de/de/themen/umwelt-infrastruktur/umweltpolitik/18324.htm.

Guo, L. (2007), *China's New Cadre Evaluation System*, EAI Background Brief No. 324 (Singapore: East Asian Institute).

Gupta, J. (2009), 'The multi-level governance challenge of climate change,' *Journal of Integrative Environmental Sciences* 4, 131–7.

Haas, P. M. (2004), 'Addressing the Global Governance Deficit,' *Global Environmental Politics* 4, 1–15.

Haites, E. and Yamin, F. (2000), 'The clean development mechanism: Proposals for its operation and governance,' *Global Environmental Change* 10, 27–45.

Hall, R. B. and Biersteker, T. J. (eds) (2002), *The emergence of private authority in global governance* (Cambridge: Cambridge University Press).

Hanely, N., Shogren, J. F. and White, B. (1997), *Environmental Economics: In Theory and Practice* (Basingstoke: Palgrave Macmillan).

Harris, P. G. (ed.) (2005), *Confronting Environmental Change in East and Southeast Asia: Eco-Politics, Foreign Policy, and Sustainable Development* (New York: United Nations University Press).

Hassett, K. A. and Metcalf, G. E. (1996), 'Can Irreversibility Explain the Slow Diffusion of Energy Saving Technologies?,' *Energy Policy* 24, 7–8.

Haufler, V. (1995), 'Crossing the Boundary between Public and Private: International Regimes and Non-State Actors,' in Rittberger, V. (ed.), *Regime Theory and International Relations* (Oxford: Clarendon Press).

Haya, B. (2007), *Failed Mechanism. How the CDM is subsidizing hydro developers and harming the Kyoto Protocol* (Berkeley, CA: International Rivers).

He, G. and Morse, R. K. (2010), *Making carbon offsets work in the developing world: lessons from the Chinese wind controversy,* Program on Energy and Sustainable Development Working Paper No. 90 (Stanford, CA: Stanford University).

Heggelund, G. and Ying, S. (2008), *CDM Development in China: China's Climate Change Policy – Through the Prism of Energy Policy* (Beijing and Lysaker: Fridtjof Nansens Institute).

Heilmann, S. (2008), 'Experimentation under Hierarchy: Policy Experiments in the Reorganization of China's State Sector, 1978–2008,' Center for International Development Working paper no. 172.

Ho, P. and Vermeer, E. B. (eds) (2006), *China's Limits to Growth: Prospects for Greening State and Society* (Oxford: Blackwell Publishers).

Höhne, N., Michelsen, C., Moltmann, S., Ott, H. E., Sterk, W., Thomas, S., Watanabe, R., Lechtenböhmer, S. and Schallböck, K. O. (2008), *Proposals for contributions of emerging economies to the climate regime under the UNFCCC post 2012,* Research Report 364 01 003 (Dessau: Umweltbundesamt).

Holzinger, K., Jörgens, H. and Knill, C. (eds) (2007), *Transfer, Diffusion und Konvergenz von Politiken* (Wiesbaden: VS-Verlag).

Hu, A. G. Z. and Jefferson, G. H. (2008), 'Science and Technology in China,' in Brandt, L. and Rawski, T. G. (eds), *China's Great Economic Transformation* Cambridge (Cambridge: Cambridge University Press).

Hu, X. and Zheng, S. (2005), *CDM Implementation in China* (Beijing: Energy Research Institute).

Huang, Y. (2002), *Selling China: Foreign Direct Investment During the Reform Era* (Cambridge, MA: Cambridge University Press).

Huber, M. R., Ruitenbeek, J. and Da Motta, R. S. (1998), *Market based Instruments for Environmental Policymaking in Latin America and the Caribbean. Lessons from Eleven Countries,* World Bank Discussion Paper (Washington, DC: World Bank).

Hunan CDM Center (2008), *Our center participates in the work of drafting the Hunan Provincial Climage Change Program [Wo Zhongxin Yu HunanSheng Yingdui Qihoubianhua Fangbianzhi Gongzuo],* at www.hncdm.com/Article/Html/6822. html.

IEA (2000), *Experience Curves for Energy Technology Policy* (Paris: International Energy Agency).

IEA (2007), *World Energy Outlook 2007* (Paris: International Energy Agency).

IEA (2010a), *Energy Technology Perspectives 2010. Scenarios and Strategies to 2050* (Paris: International Energy Agency).

IEA (2010b), *World Energy Outlook 2010* (Paris: International Energy Agency).

ILO (1997), *Business Development Services for SMEs: a Preliminary Guideline for Donor-Funded Interventions. A report to the Donor Committee on Small Enterprise Development* (Geneva: International Labor Organization).

IPCC (2001), *Climate Change 2001: Impacts, Adaptation and Vulnerability* (New York: Intergovernmental Panel on Climate Change).

Jacob, K., Beise, M., Blazejczak, J., Edler, D., Haum, R., Jänicke, M., Löw, T., Petschow, U. and Rennings, K. (2005), *Lead Markets for Environmental Innovations* (Heidelberg: Physica-Verlag and Zentrum für Europäische Wirtschaftsforschung GmbH).

Jacobs, B. J. (1979), 'A Preliminary Model of Particularistic Ties in Chinese Political Alliances: Kan-ch'ing and Kuan-hsi in a Rural Taiwanese Township,' *China Quarterly* 78, 238–73.

Jaffe, A. B. and Lerner, J. (2004), *Innovation and its discontents: how our broken patent system is endangering innovation and progress and what to do about it* (Princeton, NJ: Princeton University Press).

Jaffe, A. B., Newell, R. G. and Stavins, R. N. (2002), 'Environmental Policy and Technological Change,' *Environmental and Resource Economics* 22, 41–69.

Jaffe, A. B. and Stavins, R. N. (1995), 'Dynamic Incentives of Environmental Regulations: The Effects of Alternative Policy Instruments on Technology Diffusion,' *Journal of Environmental Economics and Management* 29, 43–63.

Jagers, S. C. and Stripple, J. (2003), 'Climate governance beyond the state,' *Global governance. A review of multilateralism and international organizations* 9, 385–99.

Jahiel, A. R. (1997), 'The contradictory impact of reform on environmental protection in China,' *China Quarterly* 149, 81–103.

Jahiel, A. R. (1998), 'The organization of environmental protection in China,' *China Quarterly* 156, 757–87.

Jänicke, M. (1990), *State-Failure. The Impotence of Politics in Industrial Society* (Cambridge: Polity Press).

Jänicke, M., Mez, L., Bechsgaard, P. and Klemmensen, B. (1999), 'Innovative Effects of Sector Specific Regulation Patterns. The Example of Energy Efficient Refrigerators in Denmark,' in Klemmer, P. (ed.), *Innovation and the Environment. Case Studies on the Adaptive Behavior in Society and the Economy* (Berlin: Analytica Verlagsgesellschaft).

Jänicke, M. and Weidner, H. (eds) (2002), *Capacity Building in National Environmental Policy. A Comparative Study of 17 Countries* (Berlin: Springer).

Jefferson, G. H. (2006), 'High Tech Regions in China: Pathways to Technology Diffusion or Centers of Increasing Technology Concentration?,' paper presented at the PRIE-ITRI workshop on High Tech Regions: Sustainability and Reinvention, November 13–14, Beijing.

Jia, H. (2006), 'Climate change, science communication and public engagement,' paper presented at the 9th International Conference on Public Communication of Science and Technology, May 17–20, Seoul, South Korea.

Jia, H. and Lin, Z. (eds) (1994), *Changing Central-Local Relations in China: Reform and State Capacity* (Boulder, CO: Westview Press).

Kaul, I., Conceição, P., Goulven, K. L. and Mendoza, R. U. (2005), 'How to improve the provision of global public goods,' in Kaul, I. (ed.), *Providing global public goods: managing globalization* (New York: United Nations Development Programme).

Kaul, I. (2008), 'Providing (Contested) Global Public Goods,' in Rittberger, V. and Nettesheim, M. (eds), *Authority in the Global Political Economy* (Basingstoke: Palgrave Macmillan).

Keck, M. E. and Sikkink, K. (1998), *Activists Beyond Borders: Advocacy Networks in International Politics* (Ithaca, NY: Cornell University Press).

Keck, O. (1987), 'The Information Dilemma: Private Information as a Cause of Transaction Failure in Markets, Regulation, Hierarchy, and Politics,' *Journal of Conflict Resolution* 31, 139–63.

Keefer, P. and Knack, S. (2005), 'Social Capital, Social Norms and the New Institutional Economics,' in Ménard, C. and Shirley, M. M. (eds), *Handbook of New Institutional Economics* (Dordrecht, The Netherlands: Springer).

Kemp, R. (1997), *Environmental Policy and Technical Change: A Comparison of the Technological Impact of Policy Instruments* (Cheltenham: Edward Elgar Publishers).

Kogut, B. (1993), 'Introduction,' in Kogut, B. (ed.), *Country competitiveness: Technology and the organization of work* (Oxford: Oxford University Press).

Kooiman, J. (2003), *Governing as governance* (London: Sage Publications).

Krugman, P. (1994), 'The Myth of Asia's Miracle,' *Foreign Affairs* 73, 62–78.

Kwong, J. (1997), *The Political Economy of Corruption in China* (New York: M.E. Sharpe).

Lahtinen, A. (2005), *China's Western Region Development Strategy and its Impact on Qinghai Province* (Helsinki: University of Helsinki).

Lai, H. H. (2002), 'China's Western Development Program: Its Rationale, Implementation, and Prospects,' *Modern China* 28, 432–66.

Lai, H. H. (ed.) (2006), *Reform and the Non-State Economy in China. The Political Economy of Liberalization Strategies* (Basingstoke: Palgrave Macmillan).

Lam, T.-C. and Perry, J. L. (2001), 'Service Organizations in China. Reforms and Institutional Constraints,' *Review of Policy Research* 18, 15–35.

Lampton, D. M. (1987), 'The Implementation Problem in Post-Mao China,' in Lampton, D. M. (ed.), *Policy Implementation in Post-Mao China* (Berkeley, University of California Press).

Lan, X., Simonis, U. E., Dudek, D. J. and et al. (eds) (2006), *Environmental governance in China* (Berlin: Wissenschaftszentrum Berlin für Sozialforschung).

Lane, D. and Myant, M. (2007), *Varities of Capitalism in Post-Communist Countries* (Basingstoke: Palgrave Macmillan).

Larson, B. and Bluffstone, R. (1997), 'Controlling Pollution in Transition Economies: Introduction to the Book and Overview of Economic Concepts,' in Bluffstone, R. and Larson, B. (eds), *Controlling Pollution in Transition Economies: Theories and Methods* (Lyme: Edward Elgar).

Lau, R. W. K. (1999), 'The 15th congress of the Chinese Communist Party: Milestone in China's privatization,' *Capital & Class* 23: 2, 51–87.

Lee, P. N.-S. and Lo, C. W.-H. (2001), 'Remaking China's Public Management: Problem Areas and Analytical Perspectives,' in Lee, P. N.-S. and Lo, C. W.-H. (eds), *Remaking China's Public Management* (Westport, CT: Quorum Books).

Lee, P. N. S. (1987), *Industrial management and economic reform in China, 1949–1984* (Hong Kong: Oxford University Press).

Lemos, M. C. and Agrawal, A. (2006), 'Environmental Governance,' *Annual Review of Environmental Resources* 31, 297–325.

Leutner, M. (2007), 'Kooperationsnetze und Akteure im semi-kolonialen China, 1860–1911,' in Risse, T. and Lehmkuhl, U. (eds), *Regieren ohne Staat? Governance in Räumen begrenzter Staatlichkeit* (Baden-Baden: Nomos).

Levitsky, J. (2000), 'Summary Report,' in Levitsky, J. (ed.), *Business Development Services. A review of international experiences* (London: Intermediate Technology Publications).

Lieberthal, K. (1997), 'China's Governing System and its impact on environmental policy implementation,' *China Environment Series* 1, 3–8.

Lipsky, M. (1983), *Street-level Bureaucracy: Dilemmas of the Individual in Public Services* (New York: Russell Sage Foundation).

Liu, X. and White, S. (2001), 'Comparing innovation systems: a framework and application to China's transitional context,' *Research Policy* 30, 1,091–114.

Lo, C. W.-H., Lo, J. M.-K. and Cheung, K.-C. (2001a), 'Institutional Reform in Environmental Governance in the People's Republic of China: Service Organizations as an Alternative for Administrative Enhancement,' *Review of Policy Research* 18, 36–57.

Lo, C. W.-H., Lo, J. M.-K. and Cheung, K.-C. (2001b), 'Service Organizations in the Environmental Governance System of the People's Republic of China,' in LEE, P. N.-S. and LO, C. W.-H. (eds), *Remaking China's Public Management* (Westport, CT: Quorum Books).

Lo, C. W.-H. and Tang, S.-Y. (2006), 'Institutional Reform, Economic Changes, and Local Environmental Management in China: the Case of Guangdong Province,' *Environmental Politics* 15, 190–210.

Lo, C. W.-H. and Tang, S.-Y. (2007), 'Institutional Reform, Economic Changes, and Local Environmental Management in China: the Case of Guangdong Province,' in Mol, A. P. J. and Carter, N. T. (eds), *Environmental Governance in China* (Abingdon and New York: Routledge).

Lothspeich, R. and Chen, A. (1997), 'Environmental protection in the People's Republic of China,' *Journal of Contemporary China* 6, 33–60.

Lundvall, B.-A. (ed.) (1992), *National Systems of Innovation – Towards a Theory of Innovation and Interactive Learning* (London: Pinter Publishers).

Lutsey, N. and Sperling, D. (2008), 'America's bottom-up climate change mitigation policy,' *Energy Policy* 36, 673–85.

Ma, X. and Ortolano, L. (2000), *Environmental Regulation in China: Institutions, Enforcement, and Compliance* (Lanham, MD: Rowman and Littlefield).

Majone, G. (2005), 'Strategy and Structure of the Political Economy of Agency Independence and Accountability,' in OECD (ed.), *Designing Independent and Accountable Regulatory Authorities for High Quality Regulation* (Paris: OECD), 126–55.

Maleki, A. (1991), *Technology and Economic Development* (New York: John Wiley).

Malmberg, A. and Power, D. (2005), 'On the role of global demand in local innovation processes,' in Fuchs, G. and Shapira, P. (eds), *Rethinking Regional Innovation and Change: Path Dependency or Regional Breakthrough* (New York: Springer).

Marr, A. (2004), *Institutional approaches to the delivery of business development: A review of recent literature, NRI Report No. 2771* (Kent: Natural Resource Institute/ DFID/World Bank).

Martinot, E. (1998), *Monitoring and Evaluation of Market Development in World Bank–GEF Climate Change Projects: Framework and Guidelines*, Climate Change Series (Washington, DC: Global Environmental Facility).

Mason, M. (2005), *The New Accountability* (London, Earthscan).

McDonald, A. and Schrattenholzer, L. (2002), 'Learning Curves and Technology Assessment,' *International Journal of Technology Management* 23, 718–45.

McElroy, M. B., Nielsen, C. P. and Lydon, P. (eds) (1998), *Energizing China: Reconciling Environmental Protection and Economic Growth* (Cambridge, MA: Harvard University Press).

McGregor, R. (2006), *China drops 'green' GDP index plan*, at www.ft.com/ cms/s/844a3cee-dfc2–11da-afe4–0000779e2340,Authorised=false.html?_i_ location=http%3A%2F%2Fwww.ft.com%2Fcms%2Fs%2F0%2F844a3cee-dfc2–11da-afe4–0000779e2340.html&_i_referer=http%3A%2F%2Fsearch.ft.c om%2Fsearch%3FqueryText%3DChina%2Bdrops%2B%2527green%2527%2B GDP%2Bindex%2Bplan.

Meyer, C. (1992), 'A Step Back as Donors Shift Institution Building from the Public to the "Private Sector",' *World Development* 20, 1,115–26.

Meyer, C. A. (1995), 'Opportunism and NGOs: Entrepreneurship and Green North-South Transfers,' *World Development* 23, 1,277–89.

Meyers, S. and Marquis, D. G. (1969), *Successful Industrial Innovation* (Washington, DC: National Science Foundation).

Michaelowa, A. (1996), 'Incentive aspects of activities implemented jointly,' *Mitigation and Adaptation Strategies for Global Change* 1: 1, 95–108.

Michaelowa, A., Jusen, A., Krause, K., Grimm, B. and Koch, T. (2000), *CDM Projects in China's Energy Supply and Demand Sectors-Opportunities and Barriers*, HWWA Discussion Paper (Hamburg: Hamburg Institute of International Economics).

Michaelowa, A. and Stronzik, M. (2002), *Transaction costs of the Kyoto Mechanisms*, HWWA Discussion Paper (Hamburg: Hamburg Institute of International Economics).

Miehlbradt, A. O. and McVay, M. (2003), *Developing commercial markets for business development services* (Geneva: International Labour Organization).

Miles, I., Kastrinos, N., Flanagan, K., Bilderbeek, R., Hertog, B., Huntink, W. and Bouman, M. (1995), *Knowledge-Intensive Business Services: Users, Carriers and Sources of Innovation*, EIMS Publication No. 15 (Luxembourg: European Innovation Monitoring System (EIMS)).

Mill, J. S. (1973), 'Of the Four Methods of Experimental Inquiry,' in Robson, J. M. (ed.), *A System of Logic* (Toronto: University of Toronto).

Minogue, M. (2001), 'Should flawed models of public management be exported? Issues and practices,' in McCourt, W. and Minogue, M. (eds), *The Internationalization of Public Management. Reinventing the Third World State* (Cheltenham: Edward Elgar).

Mol, A. P. J. and Carter, N. T. (2006), 'China's Environmental Governance in Transition,' *Environmental Politics* 15, 149–70.

Mol, A. P. J. and Carter, N. T. (2007), 'China's Environmental Governance in Transition,' in Mol, A. P. J. and Carter, N. T. (eds), *Environmental Governance in China* (Abingdon and New York: Routledge).

Moore, M., Stewart, S. and Hudock, A. (1995), *Institution Building as a Development Assistance Method. A Review of Literature and Ideas* (Stockholm: Swedish International Development Authority (SIDA)).

Morgenstern, R. D. and Al-Jurf, S. (1999), 'Can Free Information Really Accelerate Technology Diffusion?' *Technological Forecasting and Social Change* 61, 13–24.

Morton, K. (2005), *International Aid and China's Environment. Taming the Yellow Dragon* (London and New York: Routledge).

Mowery and Rosenberg, N. (1979), 'The Influence of Market Demand upon Innovation: A Critical Review of Some Recent Empirical Studies,' *Research Policy* 8, 102–53.

Murphy, L. and Edwards, P. (2003), *Bridging the Valley of Death: Transitioning from Public to Private Sector Financing* (Golden, CO: National Renewable Energy Laboratory).

National Bureau of Statistics (2008), *Local annual statistical report for Hunan 2007 [Difang Niandu Tongji Gongbao Hunan]* (Beijing: National Bureau of Statistics of China).

National Bureau of Statistics (2005), *China Statistical Yearbook 2005 [Zhongguo Tongji Nianjian]* (Beijing: National Bureau of Statistics of China).

Naugton, B. (1995), *Growing out of the Plan: Chinese Economic Reform, 1978–1993* (New York: Cambridge University Press).

NBSC (2010), *Communiqué on Energy Consumption per Unit of GDP by Region in 2009* (Beijing: National Bureau of Statistics of China).

NBSC (2009), *Communiqué on Major Data of the Second National Economic Census (No.1)* (Beijing: National Bureau of Statistics of China).

NDRC (2004), *Interim Measures for the Management of CDM Project Activities* (Beijing: National Development and Reform Commission).

NDRC (2007), *China's National Climate Change Programme* (Beijing: National Development and Reform Commission).

NDRC (2010), Letter including autonomous domestic mitigation actions, at unfccc.int/home/items/5265.php.

NDRC, MOST and MOF (2005), *Measures for Operation and Management of Clean Development Mechanism Projects in China* (Beijing: National Development and Reform Commission).

Nee, V. (1992), 'Organizational Dynamics of Market Transition: Hybrid Forms, Property Rights, and Mixed Economy in China,' *Administrative Science Quarterly* 37, 1–27.

Nelson, R. and Rosenberg, N. (1993), 'Technical innovation and national systems,' in Nelson, R. (ed.), *National Systems of Innovation: A Comparative Study* (Oxford: Oxford University Press).

Netherlands Environmental Assessment Agency (2008), *China contributing two thirds to increase in CO_2 emissions*, at www.pbl.nl.

Newell, P. (2008), 'Civil society, corporate accountability and the politics of climate change,' *Global Environmental Politics* 8, 124–55.

Newell, S., Huang, J. and Tansley, C. (2006), 'ERP Implementation: A Knowledge Integration Challenge for the Project Team,' *Knowledge and Process Management* 31, 1–12.

Ningxia DRC (2006), *The explanation of the 11th Five-Year-Plan of the Autonomous Region of Ningxia [Guanyu Ningxia Zizhiquguomin Jingjiheshehuifazhan Dishiyige Wunian Guihua Gangyao De Shuoming]* (Yinchuan: Ningxia Development and Reform Commission).

Ningxia Science and Technology Department (2006), *Ningxia's 15th Plan for Scientific Development: Report on the Strategic Development [Ningxia Shiyiwu Kejifazhan Guihuazhanlüe Yanjiubaogao]* (Yinchuan: Ningxia Science and Technology Department).

Njoroge, E. (2007), 'CDM beyond the numbers,' Presentation at the Global Carbon Market Fair and Conference, May 2–4, Cologne, Germany.

Nondek, L. and Niederberger, A. A. (2005), 'Statistical analysis of CDM capacity-building needs,' *Climate Policy* 4, 249–68.

Nunberg, B. (1998), *The State after Communism* (Washington, DC: World Bank).

O'Connor, D. (1996), *Applying Economic Instruments in Developing Countries: From Theory to Implementation*, OECD Special Papers, (Paris, OECD).

Obertheitmann, A. (2005), 'Approaches towards Sustainable Development in China,' *Journal of Current Chinese Affairs – China aktuell* 34, 41–64.

OECD (1995a), *Boosting Business Advisory Services* (Paris: OECD).

OECD (1995b), *DAC Orientations on Participatory Development and Good Governance* (Paris: OECD).

OECD (1995c), *Developing Environmental Capacity. A Framework for Donor Envolvement* (Paris: OECD).

OECD (2000), *Donor Support for Institutional Capacity Development in Environment: Lessons Learned* (Paris: OECD).

OECD (2001), *Environmental Priorities for China's Sustainable Development* (Paris: OECD).

OECD (2005a), *Governance in China* (Paris: OECD).

OECD (2005b), *Governance of Innovation Systems. Synthesis Report* (Paris: OECD).

OECD (2005c), *OECD Economic Surveys: China* (Paris: OECD).

OECD (2005d), *Part I – Mobilising Private Investment for Development: Policy Lessons on the Role of ODA,* The DAC Journal (Paris: OECD).

OECD (2006a), *Innovation and Knowledge-Intensive Service Activities* (Paris: OECD).

OECD (2006b), *Promoting Pro-Poor Growth: Private Sector Development* (Paris: OECD).

OECD (2007a), *Business for Development. Fostering the Private Sector. A Development Centre Perspective* (Paris: OECD).

OECD (2007b), *OECD Environmental Performance Reviews: China* (Paris: OECD).

Oi, J. (1994), 'Cadre networks, information diffusion, and market production in coastal China,' paper prepared for the World Bank Project on 'Explaining Growth: Chinese Coastal Provinces and Mexican Maquiladoras' (Washington, DC: World Bank).

Oi, J. C. (1992), 'Fiscal reform and the economic foundations of local state corporatism in China,' *World Politics* 45, 99–126.

Oi, J. C. (1995), 'The role of the local state in China's transitional economy,' *China Quarterly,* 144, 1,132–49.

Oi, J. C. (1996), 'The Role of the Local State in China's Transitional Economy,' in Walder, A. G. (ed.), *China's Transitional Economy* (Oxford and New York: Oxford University Press).

Oi, J. C. (1998), 'The evolution of local state corporatism,' in Walder, A. G. (ed.), *Zouping in Transition: The Process of Reform in Rural North China* (Cambridge, MA: Harvard University Press).

Olson, M. (1971), *The Logic of Collective Action: Public Goods and the Theory of Groups* (Cambridge, MA: Harvard University Press).

Osborne, D. and Gaebler, T. (1992), *Reinventing Government. How the Entrepreneurial Spirit is Transforming the Public Sector* (Reading, MA: Addison-Wesley).

Osborne, S. P. and McLaughlin, K. (2002), 'From public administration to public governance: public management and public services in the twenty-first century,' in Osborne, S. P. and McLaughlin, K. (eds), *Public Management: Critical Perspectives* (London: Routledge).

Ostrom, E. (1990), *Governing the commons: the evolution of institutions for collective action* (Cambridge: Cambridge University Press).

Ouchi, F. (2001), *Twinning as a Method for Institutional Development: A Desk Review,* WBI Evaluation Studies (Washington, DC: World Bank Institute).

Pan, J. (2003), 'The Impact of a mitigation of climatic change on the economies and politics of different regions,' *World Economics and Politics* 6, 66–71.

Panayotou, T. (1998), *Instruments of Change: Motivating and Financing Sustainable Development* (London: Earthscan).

Park, S. H. and Luo, Y. (2001), 'Guanxi and Organizational Dynamics: Organizational Networking in Chinese Firms,' *Strategic Management Journal* 22, 455–77.

Pattberg, P. and Stripple, J. (2008), 'Beyond the public and private divide: remapping transnational climate governance in the 21st century,' *International Environmental Agreements: Politics, Law and Economics* 8, 367–88.

Pearce, D., Markandya, A. and Barbier, E. (1989), *Blueprint for a Green Economy* (London: Earthscan).

Pearson, M. M. (1997), *China's New Business Elite. The Political Consequences of Economic Reform* (Berkeley: University of California Press).

Pfeffer, J. (1972), 'Merger as a response to organizational interdependence,' *Administrative Science Quarterly* 25, 129–41.

Pfeffer, J. and Nowak, P. (1976), 'Joint ventures and interorganizational dependence,' *Administrative Science Quarterly* 21, 398–418.

Pfeffer, J. and Salancik, G. R. (1978), *The External Control of Organisations: A Resource Dependence Perspective* (New York: Harper & Row).

Pierre, J. and Peters, B. G. (2000), *Governance, Politics and the State* (Basingstoke: Macmillan).

Pindyck, R. (1991), 'Irreversibility, Uncertainty, and Investment,' *Journal of Economic Literature* 29, 1,110–52.

Policy Solutions and CSEND (2005), *UNCTAD Clean Development Mechanism Training for Investment Promotion Agencies* (Washington, DC: Policy Solutions and Centre for Socio-Eco-Nomic Development (CSEND)).

Pollitt, C., Talbot, C., Caulfield, J. and Smullen, A. (2004), *Agencies: How Governments do Things Through Semi-Autonomous Organisations* (Basingstoke: Palgrave MacMillan).

Potier, M. (1995), 'The Experience of OECD Countries in their Domestic Use of Economic Instruments for Environmental Management,' in Moldan, B. (ed.), *Economic Instruments for Sustainable Development* (Prague: Ministry of Environment of the Czech Republic).

Powell, B. (ed.) (2008), *Making Poor Nations Rich. Entrepreneurship and the Process of Economic Development* (Stanford, CA: Stanford University Press).

PriceWaterhouseCoopers (2000), *A Business View on Key Issues Relating to Kyoto Mechanisms* (London: PriceWaterhouseCoopers).

Puhl, A. (1998), *Status of research on project baselines under the UNFCCC and the Kyoto Protocol* (Paris: Annex I Expert Group on the UNFCCC/OECD).

Qi, Y., Ma, L., Zhang, H. and Li, H. (2008), 'Translating a Global Issue Into Local Priority. China's Local Government Response to Climate Change,' *Journal of Environment and Development* 17, 379–400.

Reuters (2009), *China minister rejects U.S. pollution duty idea*, at www.reuters.com/article/idUSN18469068.

Rhodes, R. (1996), 'The New Governance: Governing without Government,' *Political Studies* 44, 652–67.

Risse, T. and Lehmkuhl, U. (2006), *Governance in Areas of Limited Statehood – New Modes of Governance?*, SFB Governance Working Paper Series (Berlin: SFB 700).

Roberts, D. (2008), 'Thinking globally, acting locally. Institutionalizing climate change at the local government level in Durban, South Africa,' *Environment and Urbanization* 20, 521–37.

Rogers, E. M. (2003), *Diffusion of Innovations* (New York: Free Press).

Rosenau, J. N. (1992), 'Governance, Order, and Change in World Politics,' in Rosenau, J. N. and Czempiel, E.-O. (eds), *Governance without Government. Order and Change in World Politics* (Cambridge, MA, Cambridge University).

Ross, L. (1998), 'China: Environmental Protection, Domestic Policy Trends, Patterns of Participation in Regimes and Compliance with International Norms,' *China Quarterly* 156, 809–35.

Rüb, F. W. (2002), 'Hybride Regime – Politikwissenschaftliches Chamäleon oder neuer Regimetypus? Demokratietheoretische Überlegungen zum neuen Pessimismus in der Transitologie,' in Bendel, P., Croissant, A., Rüb, F. W. (eds), *Zwischen Demokratie und Diktatur. Zur Konzeption und Empirie demokratischer Grauzonen* (Opladen: Leske & Budrich).

Sabel, C. F. (1993), 'Studied trust: Building new forms of cooperation in a volatile economy,' *Human Relations* 46, 1,133–70.

Saich, T. (2000), 'Negotiating the State: The Development of Social Organizations in China,' *China Quarterly* 161, 124–41.

Salamon, L. M. (1987), 'Of Market Failure, Voluntary Failure, and Third-Party Government: Towards a Theory of Government–Non-profit Relations in the Modern Welfare State,' *Journal of Voluntary Action Research*, 29–46.

Savas, E. S. (1979), *Privatization. The Key to Better Government* (Chatham: New Jersey).

Schneider, L. (2007), *Is the CDM fulfilling its environmental and sustainable development objectives? An evaluation of the CDM and options for improvement*, Report prepared for the WWF (Berlin: Öko-Institut).

Schreurs, M. A. (2006), 'Perspectives on Environmental Governance,' in Lan, X., Simonins, U. E., Dudek, D. J. et al. (eds), *Environmental governance in China* (Berlin: Wissenschaftszentrum Berlin für Sozialforschung).

Schreurs, M. A. (2008), 'From the Bottom Up: Local and Subnational Climate Change Politics,' *Journal of Environment and Development* 17, 343–55.

Schröder, M. (2008), 'The construction of China's climate politics: transnational NGOs and the spiral model of International Relations,' *Cambridge Review of International Affairs* 21, 505–25.

Schroeder, M. (2009), 'Varieties of carbon governance: Utilising the Clean Development Mechanism for Chinese priorities,' *Journal of Environment and Development* 18: 4, 371–94.

Schubert, G. and Tetzlaff, R. (1998), 'Erfolgreiche und blockierte Demokratisierung in der postkolonialen und postsozialistischen Weltgesellschaft – Eine Einführung,' in Schubert, G. and Tetzlaff, R. (eds), *Blockierte Demokratien in der Dritten Welt* (Opladen: Leske & Budrich).

Schumpeter, J. A. (1912/1961), *The Theory of Economic Development* (Cambridge, MA: Harvard University Press).

Schumpeter, J. A. (1939), *Business Cycles* (New York: McGraw-Hill).

Schuppert, G. F. (2008), *Von Ko-Produktion von Staatlichkeit zur Co-Performance of Governance. Eine Skizze zu kooperativen Governance-Strukturen von den Condottieri der Renaissance bis zu Public Private Partnerships*, SFB-Governance Working Paper Series Nr. 12 (Berlin: SFB 700).

Schwartz, J. (2000), 'Understanding Enforcement: Environment and State Capacity in China,' *Sinosphere* 10, 5–18.

Shama, A. (1983), 'Energy Conservation in US Buildings, Solving the High Potential/Low Adoption Paradox from a Behavioral Perspective,' *Energy Policy* 11, 148–68.

Shapiro, C. and Varian, H. (1999), *Information Rules* (Boston: Harvard Business School Press).

Shapiro, J. (2001), *Mao's War against Nature: Politics and the Environment in Revolutionary China* (Cambridge: Cambridge University Press).

Shiu, G. M. C. (1997), *Searching for Stones to Cross the River: Micro-experimental approach to Economic Reform-Lessons from China*, PhD thesis at George Mason University.

Shogren, J. F. (2007), 'The political economy of environmental governace in the United States,' in Breton, A., Brosio, G., Dalmazzone, S. and Garrone, G. (eds), *Environmental Governance and Decentralisation* (Cheltenham: Edward Elgar).

Shuwen, J. (2004), 'Assessing the dragon's choice: the use of market-based instruments in Chinese environmental policy,' *Georgetown International Environmental Law Review* 16, 617–55.

Sicular, T. (1996), 'Redefining State, Plan and Market: China's Reforms in Agricultural Commerce,' in Walder, A. G. (ed.), *China's Transitional Economy* (Oxford: Oxford University Press).

Silverberg, G. (1991), 'Adoption ad Diffusion of Technology as a Collective Evolutionary Process,' *Technological Forecasting and Social Change* 39, 67–80.

Simon, H. (1972), 'Theories of Bounded Rationality,' in McGuire, C. B. and Radner, R. (eds), *Decision and Organiszation* (New York: American Elseview).

Sinkule, B. J. and Ortolano, L. (1995), *Implementing Environmental Policy in China* (Westport, CT: Praeger Publishers).

Sippel, M. and Jenssen, T. (2010), *What about Local Climate Governance? A Review of Promise and Problems*, MPRA Paper No. 20987 (Munich: Munich University).

Smil, V. (2003), *China's Past, China's Future: Energy, Food, Environment* (New York: Routledge).

Smith, K. (2001), 'Human Resources, Mobility and the Systems Approach to Innovation,' in OECD (ed.), *Innovative People: Mobility of Skilled Personnel in National Innovation Systems* (Paris: OECD).

Snehota, I. (2004), 'Perspectives and theories of market,' in Hakasson, H., Harrison, D. and Waluszewski, A. (eds), *Rethinking Marketing. Developing a new understanding of markets* (Chichester: John Wiley & Sons).

Solinger, D. J. (1992), 'Urban Entrepreneurs and the State: The Merger of State and Society,' in Rosenbaum, A. L. (ed.), *State and Society in China. The Consequences of Reform* (Boulder, CO: Westview Press).

Spofford, W., Ma, X., Ji, Z. and Smith, K. (1996), *Assessment of the Regulatory Framework for Water Pollution Control in the Xiaoqing River Basin: A Case Study of Jinan Municipality* (Washington, DC: World Bank).

Stark, D. (1996), 'Coexisting Organizational Forms in Hungary's Emerging Mixed Economy,' in Nee, V. and Stark, D. (eds), *Remaking the Economic Institutions of Socialism: China and Eastern Europe* (Stanford: Stanford University Press).

State Council (1998), *Tentative Regulation on Public Service Units Registration*, October 25, 1998 edition (Beijing: State Council of the PR China).

Stavins, R. N. (2000), 'Market-based environmental policies,' in P. R. Portney and R. N. Stavins (eds), *Public Policies for Environmental Protection*, 2nd edition (Washington, DC: Resources for the Future).

Stavins, R. N. (2001), *Experience with Market-Based Environmental Policy Instruments*, Discussion Paper 01–58 (Washington, DC: Resources for the Future).

Steel, W., Tanburn, J. and Hallberg, K. (2000), 'The emerging strategy for building business development service markets,' in Levitsky, J. (ed.), *Business*

Development Services: A Review of International Experience (London: Intermediate Technology Publications).

Stella, P. (1992), 'Tax Farming – A Radical Solution for Developing Country Tax Problems,' *Staff Papers – International Monetary Fund* 40: 1, 217–25.

Stern, N. (2006), *The Economics of Climate Change: The Stern Review* (Cambridge: Cambridge University Press).

Stiglitz, J. (1989), *Wither Socialism?* (Cambridge and London: MIT Press).

Stoker, G. (1998), 'Governance as Theory: Five Propositions,' *International Social Science Journal* 50, 17–28.

Stokey, N. L. (1986), 'The Dynamics of Industrywide Learning,' in Heller, W. P., Starr, R. M. and Starrett, D. A. (eds), *Equilibrium Analysis. Essays in Honor of Kenneth J. Arrow* (New York: Cambridge University Press).

Stölting, E. (2001), 'Das Öffentliche an öffentlichen Unternehmen,' in Edeling, T., Jann, W., Wagner, D. and Reichard, C. (eds), *Öffentliche Unternehmen. Entstaatlichung und Privatisierung?* (Opladen: Leske & Budrich).

Stoneman, P. and Diederen, P. (1994), 'Technology Diffusion and Public Policy,' *The Economic Journal* 104, 918–30.

Streck, C. (2007), 'The governance of the Clean Development Mechanism: the case for strength and stability,' *Environmental Liability* 2, 91–100.

Sugiyama, N. and Takeuchi, T. (2008), 'Local Policies for Climate Change in Japan,' *The Journal of Environment and Development* 17, 424–41.

Tang, S.-Y., L., Wing-Hung, C., Wing-Hung., Cheung, K.-C. and Man-Keung, J. L. (1997), 'Institutional constraints on environmental management in urban China: environmental impact assessment in Guangzhou and Shanghai,' *China Quarterly* 152, 863–74.

Tellis, G. J., Stremersch, S. and Yin, E. (2002), 'The International Takeoff of New Products: The Role of Economics, Culture, and Country Innovativeness,' *Marketing Science* 22, 188–208.

Tenev, S., Zhang, C. and Brefort, W. L. (2002), *Corporate Governance and Enterprise Reform in China. Building the institutions of modern markets* (Washington, DC: World Bank and the International Finance Corporation).

TERI (2004), *CDM in India*, Teri Report (New Delhi: The Energy and Resources Institute).

Tews, K. (2005), 'Die Diffusion umweltpolitischer Innovationen: Eckpunkte eines Analysemodells,' in Tews, K. and Jänicke, M. (eds), *Die Diffusion umwelt-politischer Innovationen im internationalen System* (Wiesbaden: VS Verlag für Sozialwissenschaften).

The Economist (2009), *Green with envy. The tension between free trade and capping emissions*, at www.cconomist.com/node/14926073.

The Heritage Foundation (2010), 2010 Index of Economic Freedom, at www.heritage.org.

Tietenberg, T. H. (1990), 'Economic instruments for environmental regulation,' *Oxford Review of Economic Policy* 6, 17–33.

Toivonen, M. (2004), *Expertise as Business: Long-term Development and Future Prospects of Knowledge-Intensive Business Services (KIBS)* (Helsinki: Helsinki University of Technology).

Tong et al. (1999), 'Civil service reform in the People's Republic of China: case studies of early implementation,' *Public Administration and Development* 19, 193–206.

Tornatzky, L. G. and Klein, K. J. (1982), 'Innovation Characteristics and Adoption-Implementation: A Meta-Analysis of Findings,' *IEEE Transactions on Engineering Management* 29, 28–45.

Tsai, K. (2001), 'Beyond banks: The local logic of informal finance and private sector development in China' paper presented at the conference 'Financial Sector Reform in China,' September 11–13, Cambridge, MA.

Tsui, A. S., Bian, Y. and Cheng, L. (2006), 'Explaining the Growth and Development of the Chinese Domestic Private Sector,' in Tsui, A. S., Bian, Y. and Cheng, L. (eds), *China's Domestic Private Firms* (Armonk, NY: M. E. Sharpe).

Tullock, G. (1967), 'The welfare cost of tariffs, monopolies and theft,' *Western Economic Journal* 5, 224–32.

Tullock, G. (1993), *Rent-seeking* (Vermont: Edward Elgar).

Turner, J. L. and Linden, E. (2007), 'China's Growing Ecological Footprint,' *China Monitor*, 7–10.

UN (1982), *Elements of Institution-building for Institutes of Public Administration and Management* (New York: United Nations, Department of Technical Cooperation for Development).

UNECE (2002), *Guidelines on Best Practice in Business Advisory, Counselling and Information Services* (New York and Geneva: United Nations Economic Commission for Europe).

UNEP (2010), *The Emissions Gap Report. Are the Copenhagen Accord pledges sufficient to limit global warming to 2° C or 1.5° C?* (Nairobi: United Nations Environmental Programme).

UNEP Risoe (2008), *Increasing access to the carbon market* (Roskilde, Denmark: UNEP).

UNEP Risoe (2009), *CDM/JI Pipeline Analysis and Database October 1st*, at www.cdmpipeline.org.

UNEP Risoe (2010), *CDM/JI Pipeline Analysis and Database December 1st*, at www.cdmpipeline.org.

UNIDO (2007), *The role of renewable energy in China* (Beijing: UNIDO Regional Office).

Van der Gaag, J. (1995), *Private and Public Initiatives: Working Together for Health and Education* (Washington, DC: World Bank).

Verheijen, A. H. G. (2000), *Administrative Capacity Development. A Race Against Time?* (The Hague: WWR Scientific Council for Government Policy).

Verheijen, A. H. G. (2003), 'Public Administration in Post-Communist States,' in Peters, B. G. and Pierre, J. (eds), *Handbook of Public Administration* (London: Sage Publications).

Vidal, J. (2009), *Ed Miliband: China tried to hijack Copenhagen climate deal*, at www.guardian.co.uk/environment/2009/dec/20/ed-miliband-china-copenhagen-summit.

Wade, R. (1990), *Governing the Market: Economic Theory and the Role of Government in East Asian Industrialisation* (Princeton: Princeton University Press).

Walder, A. (1992), 'Local Bargaining Relationships and Urban Industrial Finance,' in Lieberthal, K. G. and Lampton, D. M. (eds), *Bureaucracy, Politics, and Decision Making in Post-Mao China* (Berkeley: University of California Press).

Walder, A. G. (1995), 'Local governments as industrial firms,' *American Journal of Sociology*, 5, 262–301.

Walder, A. G. (1996), 'China's Transitional Economy: Interpreting its Significance,' in Walder, A. G. (ed.), *China's Transitional Economy* (Oxford: Oxford University Press).

Wang, H. (2001), 'Redefining Regional Development Strategies in China,' in Edgington, D. W., Fernandez, A. L. and Hoshino, C. (eds), *New Regions – Concepts, Issues, and Practices* (Westport and London: Greenwood Press).

Wang, J. (2009), 'Opening Address,' in Fan, Q., Li, K., Zeng, D. Z., Dong, Y. and Peng, R. (eds), *Innovation for Development and the Role of Government. A Perspective from the East Asia and Pacific Region* (Washington, DC: International Bank for Reconstruction and Development/World Bank).

Wang, S. and Hu, A. (1993), *Zhongguo Guojia Nengli Baogao [A Study of Chinese State Capacity]* (Shenyang, Liaoning: People Publishing House).

Wang, S. (1995), 'The rise of the regions: fiscal reform and the decline of central state capacity in China,' in Walder, A. G. (ed.), *The Waning of the Communist State: Economic Origins of Political Decline in China and Hungary* (Berkeley: University of California Press).

Wang, S. and Hu, A. (2001), *The Chinese Economy in Crisis: State Capacity and Tax Reform* (Armonk, NY: M.E. Sharpe).

Watanabe, R., Arens, C., Mersmann, F., Ott, H. E. and Sterk, W. (2008), *The Bali Roadmap to Global Climate Policy: New Horizons and Old Pitfalls* (Wuppertal: Wuppertal Institute for Climate, Environment and Energy).

Watts, J. (2009), *China's carbon emissions will peak between 2030 and 2040, says minister*, at www.guardian.co.uk/environment/2009/dec/06/china-carbon-emissions-copenhagen-climate.

Watts, J., Vidal, J., McKie, R. and Helm, T. (2009), *China blamed as anger mounts over climate deal*, at www.guardian.co.uk/environment/2009/dec/20/china-blamed-copenhagen-climate-failure.

Weinert, R. (2001), 'Die Verflüchtigung des Politischen. Gemeinwirtschaftliche, genossenschaftliche und öffentliche Wohnungsunternehmen im Vergleich,' in Edeling, T., Jann, W., Wagner, D. and Reichard, C. (eds), *Öffentliche Unternehmen. Entstaatlichung und Privatisierung?* (Opladen: Leske & Budrich).

White, G., Howell, J. and Shang, X. (1996), *In search of civil society: market reform and social changes in contemporary china* (Oxford: Clarendon Press).

White, G. and Wade, R. (eds) (1988), *Developmental States in East Asia* (New York: St. Martin's).

Willems, S. and Baumert, K. (2003), *Institutional Capacity and Climate Actions* (Paris: OECD and IEA).

Williamson, O. (1991), 'Comparative economic organization: The analysis of discrete structural alternatives,' *Administrative Science Quarterly* 36, 269–96.

Wolf, K. D. (2008), 'Emerging Patterns of Global Governance. The New Interplay Between the State, Business and Civil Society,' in Scherer, A. G. and Palazzo, G. (eds), *Handbook Of Research On Global Corporate Citizenship* (Cheltenham/UK, Northampton/MA: Edward Elgar).

Wong, C. (2004), *Toward a Future Vision of Public Service Unit Reform in China: An Analytical Framework* (Washington, DC: World Bank).

Wong, C. P. W. (1987), 'Between plan and market: The role of the local sector in post-mao China,' *Journal of Comparative Economics* 11, 385–98.

World Bank (1996), *From Plan to Market* (Washington, DC: World Bank).

World Bank (1997a), *Clear Water, Blue Skies: China's Environment in the New Century* (Washington, DC: World Bank).

World Bank (1997b), *World Development Report 1997: The State in a Changing World* (Washington, DC: World Bank).

World Bank (2007), *East Asia and Pacific Update* (Washington, DC: World Bank).

World Bank (2008), *State and Trends of the Carbon Market* (Washington, DC: World Bank).

World Bank (2010), *Launch of the Clean Development Mechanism (CDM), South–South Cooperation between China and other Developing Countries*, at web.world-bank.org/WBSITE/EXTERNAL/WBI/WBIPROGRAMS/ENRLP/EXTCARFINAS S/0,,contentMDK:21840940~pagePK:64168445~piPK:64168309~theSitePK:32 87761,00.html.

World Bank, Ministry of Science and Technology (PR China), GTZ, Federal Ministry of Economic Cooperation and Development (Germany), Swiss State Secretariat for Economic Affairs (Switzerland) (2004), *Clean development mechanisms in China: taking a proactive and sustainable approach* (Washington, DC: World Bank).

World Resources Institute (2010), Climate Analysis Indicators Tool (CAIT), Version 8.0., at cait.wri.org.

Xia, M. (2000), *The Dual Developmental State. Development strategy and institutional arrangements for China's transition* (Aldershot: Ashgate Publishing Company).

Xie, Z. (2010), *Speech at High level Segment of COP16 & CMP6*, at unfccc.int/files/meetings/cop_16/statements/application/pdf/101208_cop16_hls_china.pd.

Yang, D. L. (2004), *Remaking the Chinese Leviathan. Market transition and the politics of governance in China* (Stanford: Stanford University Press).

Yang, J. and Schreifels, J. (2003), *Implementing SO2 Emissions in China* (Paris: OECD).

Young, D. R. (1986), 'Entrepreneurship and the behavior of non-profit organizations: Elements of a theory,' in Rose-Ackerman, S. (ed.), *The Economics of Nonprofit Institutions* (New York: Oxford University Press).

Yu, T. F. (1997), 'Entrepreneurial state: the role of government in the economic development of the Asian newly industrializing economies,' *Development Policy Review* 15, 47–64.

Zhang, C. (2002), 'The Interaction of the State and the Market in a Developing Transition Economy: the Experience of China,' paper presented at the International Seminar Promoting Growth And Welfare: Structural Changes And The Role Of Institutions In Asia, April 29 – May 03, Santiago, Chile and Rio de Janeiro, Brazil.

Zhang, C., Heller, T. C. and May, M. M. (2005), 'Carbon intensity of electricity generation and CDM baseline: case studies of three Chinese provinces,' *Energy Policy* 33, 451–65.

Zhang, E. et al (1996), *Research on the Reform of the Funding System for the Service Organizations in Shanghai [Shanghai Shi Gaige Shiye Danwei Jingfei Guanli Tizhi De Yanjiu]* (Shanghai: Tongji University).

Zhang, K. (2000), 'Policies and Actions on Sustainable Development in China,' *China Environmental Science Press*, 176–86.

Zhang, W., Vertinsky, I., Ursacki, T. and Newetz, P. (1999), 'Can China be a clean tiger?: Growth strategies and environmental realities,' *Pacific Affairs* 72: 1, 23–37.

Zhang, Z. (2005), *Towards an Effective Implementation of the Clean Development Mechanism Projects in China* (Honolulu: East-West Center).

Zhang, Z. (2008), *How far can developing country commitments go in an immediate post-2012 climate regime?* (Honolulu: East–West Center).

Zheng, Y. (1999), *Discovering Chinese nationalism in China: modernization, identity, and international relations* (Cambridge: Cambridge University Press).

Zhong, Y. (2003), *Local Government and Politics in China: Challenges from Below* (Armonk, NY and London: M. E. Sharpe).

Zhou, X. (1995), 'Partial reform and the Chinese bureaucracy in the post Mao era,' *Comparative Social Studies* 28, 440–69.

Zürn, M. (1998), *Regieren jenseits des Nationalstaates. Globalisierung und Denationalisierung als Chance* (Frankfurt a.M.: Suhrkamp).

Index